Pioneers in Arts, Humanities, Science, Engineering, Practice

Volume 11

Series editor

Hans Günter Brauch, Mosbach, Germany

More information about this series at http://www.springer.com/series/15230
http://www.afes-press-books.de/html/PAHSEP.htm
http://afes-press-books.de/html/PAHSEP_Santos.htm

Lucas Melgaço · Carolyn Prouse
Editors

Milton Santos: A Pioneer in Critical Geography from the Global South

 Springer

Editors
Lucas Melgaço
Department of Criminology
Vrije Universiteit Brussel
Brussels
Belgium

Carolyn Prouse
Department of Geography
University of British Columbia
Vancouver, BC
Canada

Acknowledgement: We would like to express our gratitude to Marie-Hélène Tiercelin, Milton Santos's wife, for her generous support throughout the writing of this book.
A book website with additional information on Milton Santos and his major book covers can be accessed at: http://afes-press-books.de/html/PAHSEP_Santos.htm.

ISSN 2509-5579   ISSN 2509-5587   (electronic)
Pioneers in Arts, Humanities, Science, Engineering, Practice
ISBN 978-3-319-53825-9   ISBN 978-3-319-53826-6   (eBook)
DOI 10.1007/978-3-319-53826-6

Library of Congress Control Number: 2017932412

The cover photograph and the photograph on page iii were taken from the collection of the magazine *Jornal da USP* (University of São Paulo). Photos by Jorge Maruta, 1994. The permission to use this and other photos by this photographer in this book was granted.

Copyediting: PD Dr. Hans Günter Brauch, AFES-PRESS e.V., Mosbach, Germany

Printed on acid-free paper

This Springer imprint is published by Springer Nature
The registered company is Springer International Publishing AG
The registered company address is: Gewerbestrasse 11, 6330 Cham, Switzerland

Contents

Chapter 1
Milton Santos and the Centrality of the Periphery

Lucas Melgaço and Carolyn Prouse

Abstract This chapter introduces Brazilian Geographer Milton Santos to English-speaking academia. The text places Santos's theorizations in conversation with recent calls to expand, postcolonialize and decolonize knowledge production in urban studies: to say that Santos is a geographer from the Global South and of the periphery is to indicate the marginalization of theories produced from particular social locations. The text then offers a biography of the author, highlighting the plurality of his experiences and the diversity of contexts in which he lived during his academic career. This introductory chapter also includes a discussion of the particularities of what could be called a Miltonian epistemology of geography, and concludes by briefly presenting the chapters that comprise the book Milton Santos: A Pioneer in Geography from the Global South.

Many scholars consider the Vautrin Lud Prize the Nobel Prize for Geography. Since 1991 the award has honored geographers who have made exceptional contributions to the world of geography. Its track record of winners, however, indicates the unequal distribution of value attributed to different geographies of thought. At the time of writing 28 scholars have received this honorable distinction: seven are from the United Kingdom, six from the United States, five from France, four from Switzerland, two from Spain, one from Ireland, one from Sweden, one from Canada and one from Brazil (Festival International de Géographie 2016). In other words, only 3.6%—one person—from outside the Europe-North America axis has been awarded the prize. These numbers demonstrate that what is considered 'excellence' in geography is restricted to a very specific circle, limited not only in terms of country of origin, but also language: 53.6% of the awarded scholars are primary English speakers, 22% speak French, 7.1% German, 7.1% Spanish, 3.6% Swedish, and 3.6%—one—is a Brazilian Portuguese speaker. This Brazilian Portuguese

Lucas Melgaço, Assistant Professor, Vrije Universiteit Brussel; Email: lucas.melgaco@vub.ac.be.

Carolyn Prouse, Ph.D. Candidate, University of British Columbia; Email: carolyn.prouse@geog.ubc.ca.

© Springer International Publishing AG 2017   1
L. Melgaço and C. Prouse (eds.), *Milton Santos: A Pioneer in Critical Geography from the Global South*, Pioneers in Arts, Humanities, Science, Engineering, Practice 11, DOI 10.1007/978-3-319-53826-6_1

speaker, the only scholar from outside the 'North' to have received the Vautrin Lud Prize, is Brazilian geographer Milton Santos.

Scholars of Anglo[1] geography are recognizing and problematizing this uneven pattern of what and who 'counts' within the discipline. This book has been written at a moment when there is considerable interest in the English-speaking geography world of learning from and with spaces and places in the South. Postcolonial theory, in particular, has made reverberations throughout this discipline, and is throwing into question truths of what counts as canonical knowledge and foundational models in its subfields (see, for instance, Radcliffe 2005; Roy 2010, 2011; Wainwright 2008; Lawson 2007; Robinson 2006; McFarlane 2008). The histories of colonialism that shape geographic knowledge production have privileged the experiences of Northern cities and nations, which have become the main sites of theory production, while Southern experiences are treated as sites of ethnographic extraction (Robinson 2006) and empirical variation (Roy 2016). Santos himself anticipated this concern over thirty years ago, worrying that 'concepts formulated on the basis of data from developed countries have been indiscriminately applied to Third World countries' (Santos 1979: 6) and that geographers must 'base an historical analysis on Third World reality rather than on the assumption that all social evolution is simultane-ously comparable and complementary' (Santos 1979: 6). Eurocentrism emerges here as an epistemological problem (Roy 2014), with the North granting itself the 'per-mission to narrate' from an unseen, universal, and objective standpoint (Sundberg 2003). A postcolonial or decolonial methodology is, in part, a mapping exercise, which points to the locatedness of all geographic knowledge production and '[-places] the permission to narrate on a map' (Roy 2016: 8).

This project represents a response to this postcolonial/decolonial problem space in two major ways. First, it not only locates a permission to narrate in Brazil, but it shows that Brazilian scholars like Santos have been narrating for decades *without* permission from the English-speaking world. Milton Santos is but one figure who theorized from the situated experiences of Brazilian urbanization before postcolonial thought gained its contemporary academic traction. He did not need an Anglo academy to grant him the permission to narrate, but claimed it for himself even when invisibilized in hegemonic geographic thought. Second, Santos did not just theorize about the oppressive political economic and colonial relations that have shaped Brazil. Within his writings and practices he also, crucially, fought for and found promise in counter-rational, non-hegemonic values and knowledges placed at risk by global capital. In a moment when much academic attention was centered on glob-alization Santos highlighted the importance of place amidst global flows: 'Each place is in its own way the world' (Santos 1996a: 252) whereby globalization represents not only perversity to places in the Global South, but also new possi-bilities (Santos 2000a). By focusing on his own and his compatriots' locatedness in

---

[1]The term 'Anglo' is fraught, of course. We use it here to refer to the English-speaking world of geography—manifest most hegemonically in high-impact journals—without trying to make specific national, ethnic or racial assumptions. 'Anglo geography', then, is a synonym for 'English-language geography' in this text.

Milton Santos, 1969. Retrieved from Santos's United Nation identity card. This photograph was taken from the personal photo collection of the author and his wife, Marie-Hélène Santos, who also granted permission for publication. This and other photos can be found at: https://www. miltonsantos.com.br/

Brazil, Santos shifted the locus of theoretical enunciation (Mignolo 2000) to two different types of marginalized experiences: forms of urbanization and capital flows not captured by hegemonic Northern theory; and spaces, places and values that are increasingly under threat by these flows and forms of urbanization.

1.1   Milton Santos: A Geographer of the South?

We use the term Global South but recognize its fraught history as it often connotes backwardness, tradition and underdevelopment. Recent scholarship in Anglo geography[2] has troubled the category of the Global South by arguing that it is not a delineated space that can be neatly mapped through political economic relations or nation-state borders. It should be, in other words, a fluid and contested category (Parnell & Oldfield 2014). In this vein, the South has been characterized variously as a concept-metaphor (Roy 2015) and relation (Comaroff & Comaroff 2012). Just as Rao argues for the term 'slum', we think of the Global South 'in a normative sense to gain visibility for certain histories and the landscapes of politics and action that they imply' (Rao 2006: 228). Milton Santos is thus a geographer *of* the Global South because his ideas are peripheralized by colonial linguistic flows and relations of knowledge that privilege Anglo geography and the experiences of Northern cities and countries. We use the term Global South, then, to denote an uneven relation in academic theorizations and to point to the political imperative in increasing the visibility of scholars and ideas that have heretofore been marginalized. A major purpose of this book is to increase the visibility of Santos's theorizations to Northern, Anglo audiences, thus subverting the unequal relation between Northern and Southern knowledges such that these categories no longer make sense.[3]

The development of Milton Santos's thought itself disrupts any neat binary distinction between Northern and Southern theory. Santos did not exist in an academic bubble in Brazil; his thinking was shaped in part by theories, models and academics in the North. His time in exile in France was vital for shaping his ideas about urbanization in Brazil and other Southern cities even as he realized these French theoretical models could not fully account for his experiences. He does not completely abandon ideas and notions that have been developed in the Northern academy. Rather, he theorizes processes of urbanization from his particular situated standpoint (Mohanty 2002): being an Afro-Brazilian geographer from Bahia who worked with colleagues from Brazil to France to Tanzania to Canada. According to

---

[2]Of course, there are many different English speakers whose works are marginalized in Anglo geography for reasons of nationality, race and ethnicity. In other words, Anglo does not map neatly onto the 'North'. But we retain the usefulness of the term 'Anglo' to describe a linguistic divide that must be traversed for thinkers from other linguistic schools to be properly understood and engaged in English—the language that is hegemonic in curriculum and high-ranked journals.

[3]The provocative idea of the 'centrality of the periphery' that serves as inspiration for our title was borrowed from Santos's book Toward an Other Globalization (Santos 2000a).

Santos himself (Yázigi 1996) his most important influences in Geography were French Marxist geographers Pierre George and Jean Tricart, his Ph.D. supervisor, from whom he 'learned the rigor, the will for discipline, the obedience to projects and the joy of argument' (Silva 2002: 13). It was Brazilian physician and geographer Josué de Castro[4] whose work *Ensaios de Geografia Humana* triggered Santos's deep interest in the discipline when he was still a high school student (Vasconcelos 2001). Santos was also strongly influenced by philosophers like Sartre and Whitehead; sociologists like Durkheim and Gurtvitch; and, of course, political economists like Marx. When asked if he was a Marxist Santos affirmed:

> My interest in history, above all the history of the present, led me to always take into account the contradictory process, in a way that not being an orthodox Marxist – I am afraid of that, of an orthodox Marxism – I believe that every doctrine that does not renew itself risks of becoming a religion, a dogma, and consequently dumb down instead of clarify. Having said that, I do consider myself a Marxist, or if you prefer, a marxizing. (Tendler 2006: 6:35)

Santos theorizes processes, too, that cannot be adequately captured by any of these theories: the massive population growth of Brazilian and other Southern cities; the informal circuits of the economy; and the conflicts over knowledges and values found at the frontiers of global capitalism. Santos thus might be called what Mignolo (2000) terms a 'border thinker': someone who operates within different knowledge cultures, thinking from and along the borders of empire, whether these empires be ongoing political economic histories or epistemicides within the academy. In the following section we trace the evolution of Santos's thought through his biography by looking at some of the experiences and major works that made Milton Santos one of the most revered geographers in Brazil and Latin America. We then discuss the fraught process of translation and conclude with a summary of this volume's contributions.

1.2 The Restless Thinker

Santos was born in Brotas de Macaúbas, a small town in the interior of the state of Bahia, in the Northeast region of Brazil on 3 May 1926. He only lived there for a year, however, as his parents were elementary school teachers and had to travel to different towns in the interior of Bahia for work. As a young child Santos did not go to school but was educated at home by his parents. They taught him how to read and write and fomented an aptitude in mathematics and French (a language that would facilitate his contacts with French academia throughout his life; Grimm 2011). At the age of ten he moved to Salvador, the capital of the state of Bahia, for

---

[4]Josué de Castro (1908–1973) is also the author of two other seminal works, *The Geography of Hunger* and *The Geopolitics of Hunger*.

boarding school. As a descendent of slaves,[5] it might be easy to assume Santos had a difficult and poor childhood. However, this was not the case: Santos came from a very well educated middle-class family, which was a situation that helped him to excel as a student from his earliest years.

Santos's first degree was not in geography but in law, which he obtained in Salvador in 1948. He only sporadically worked as a lawyer. Immediately after his legal training, Santos was hired as a geography teacher at the Municipal School of Ilhéus, a medium-sized city on the southern coast of the state of Bahia, where he lived until 1953. Even before achieving a doctorate in the discipline, then, Santos had started work as a geography teacher.

Milton Santos, 1948. Graduation in bachelor of Law. This photograph was taken from the personal photo collection of the author and his wife, Marie-Hélène Santos, who also granted permission for publication. This and other photos can be found at: https://www.miltonsantos.com.br/

Santos became editor of the newspaper *A Tarde*, a prestigious local media vehicle in the city of Salvador. As a journalist he was sent to missions in African and European countries, which resulted in a dozen articles later integrated into books such as *Marianne em Preto e Branco* (Santos 1960) and *A Cidade do Países Subdesenvolvidos* (Santos 1965). In 1960 Santos was invited by President Janio Quadros to document the latter's trip to Cuba, which resulted in the series *Visita a uma Revolução* (Silva & Silva 2004).

---

[5]Brazil is a country with a high concentration of African descendants. However, very few Afro-Brazilians attend university, much less achieve the rank of professor. Santos was one of the exceptions. Despite being concerned about issues of racism and prejudice, having published about them (Santos 1996b, 2000b) and answered questions about these topics in a handful of interviews, Santos was often critical of certain strategies and approaches of black movements and always very apprehensive about speaking in the name of Afro-Brazilians.

House of the family of Milton Santos in Salvador, Brazil, 1953. This photograph was taken from the personal photo collection of the author and his wife, Marie-Hélène Santos, who also granted permission for publication. This and other photos can be found at: https://www.miltonsantos.com.br/

Santos was also very active in local politics. He was the representative of the President of the Civil House Janio Quadros in Bahia in 1961, and from 1962 to 1964 he held the position of president of the Foundation for the Economic Planning Commission of the State of Bahia (during the mandate of President João Goulart (1961–1964)). Santos was also active as a representative of geographers and served as president of the Association of Brazilian Geographers from 1963 to 1964. Despite his political activity, Santos considered himself somewhat of an 'outsider' in Brazil, not claiming any group membership—intellectual, political or otherwise (Tendler 2006: 4:55).

Santos's academic career began in 1956 when he was appointed lecturer of geography in the Catholic University of Salvador. During this period Santos also travelled to France to study for his doctorate in Geography at the University of Strasbourg, France, under the supervision of Prof. Jean Tricart. He was awarded his doctorate in 1958 for the thesis 'Le centre de la ville de Salvador. Étude de geographie urbaine'.

Santos returned to Brazil in 1959 and in 1961 he became professor at the Federal University of Bahia. He lived in Salvador until 1964, when a military coup d'état ousted Brazil's democratically elected government. Santos was arrested by the military police and spent three months in prison. Just like many arrests during the military dictatorship, the reasons for his imprisonment were unclear, but were likely linked to his proximity to left-wing President João Goulart, to his aforementioned trip to Cuba and to the progressive content of his journalistic texts (Grimm 2011). He was released from prison on the condition that he be deported. During thirteen years of exile Santos worked as an academic in different countries in Europe, North America and Africa.

Milton Santos, 1991. Photo by Oswaldo José dos Santos, from the collection of the magazine *Jornal da USP* (University of São Paulo)

In 1977 Santos returned to Brazil, and from 1983 onwards was affiliated with the University of São Paulo where he worked until he passed away on 24 June 2001 at the age of 75. Santos received the *honoris causa* distinction from twenty universities, including one in France, two in Spain, three in Argentina, one in Uruguay, and 13 in Brazil. Milton Santos was the sole author of more than 40 books, edited 14 collections, published 58 book chapters and wrote almost 300 scientific articles during his 50-year career (Santos 2001).

Santos's wife Marie-Hélene Tiercenlin and geographer Jacques Levy periodize the geographer's career in three different phases (Tiercelin dos Santos & Levy 2011). The first period, from 1948 to 1964, began with the publication of Santos's first book, *O Povoamento da Bahia: Suas Causas Econômicas* (Santos 1948), and was characterized by the geographer's strong commitment to the political and academic context of Bahia (Grimm 2011). The second period, from 1964 to 1977, was his term of exile during which Santos traveled to and worked in many different countries and at universities throughout the world: Universities of Toulouse, Bordeaux and Paris (Panthéon Sorbonne), MIT (Massachusetts Institute of Technology—Boston, USA), University of Toronto (Canada), Caracas (Venezuela), Dar-es-Salam (Tanzania) and Columbia University (New York—USA). During this period Santos achieved some repute as a scholar specializing in theories of underdevelopment and the Third

World. Scholars from the Anglo-Europe axis who happened to know Milton Santos still have fond memories of the geographer from this period of exile. It was also the phase in which Santos authored two key works: *Les Villes Du Tiers Monde* (1971) and *The Shared Space* (1979), the latter being thus far the only monograph of Santos's translated into English.

From 1977 to 2001 Santos experienced his must prolific and mature phase as a geographer and thinker. This period remains, however, his most unknown to the Anglo-world: almost none of his important works from this phase are available in English. The 1970s were years of transformation in disciplinary geography throughout the world, including Brazil. The traditional descriptive orientation of the discipline was being questioned for its (im)possibility of explaining and understanding the complexity of the world. In Brazil, the influences of a burgeoning critical and qualitative geography (mainly from France) and a quantitative and technical geography (mainly from the United States) came into direct confrontation. Santos was very active as a supporter of the so-called critical school in geography. However, more than simply reproducing and applying foreign theories, he worked on developing a Brazilian (and Global South) school of thought. In 1978 Santos published one of his most important books, *Por uma Geografia Nova* (also available in Spanish and French), in which he analyzes the different geographic schools of thought and presents his own 'method' of geography. In this book Santos argues that geographers should focus on the object of study of geography—*geographic space*—which, according to him, is simultaneously a social construction and social factor, both constructed by and constructing society:

If the organized space is also a form, an objective result of the interaction of multiple variables throughout history, its inertia, it can be said, is dynamic. By dynamic inertia I want to say that the forms are both a result and a condition of processes. (Santos 2004: 185)

Those in the North who are familiar with Milton Santos generally consider him to be a specialist in globalization, urban geography and the Third World. But this is not entirely accurate—his importance exceeds these specific topics. Santos created what he called 'a method': a set of coherent and complementary concepts and categories that together explain different spatial phenomena. According to him:

The discussion [should be] about space and not geography; this points to the importance of mastering the method. Speaking about the object without mentioning the method might just be the announcement of a problem, without, however, enunciating it. An ontological approach is essential – an interpretive effort *from within*, which both contributes to identifying the nature of space, and to finding the categories that allow the analysis of this space. (Santos 1996a: 19, emphasis in original)

In addition to his focus on method, Santos was a very creative concept theorist. He reinterpreted traditional geographic concepts and proposed new terms and neologisms.[6] His theorizations cover considerable much conceptual ground and

---

[6]The array of terms that Santos employed was so vast that it motivated Ferreira (2000) to create a glossary of the author's terms that culmuniated in a Ph.D. dissertation in the field of linguistics.

have inspired such a great number of scholars that a 'Miltonian' school of geographic thought has emerged in Latin America, particularly Brazil. This book offers
readers a chance to become acquainted with this Miltonian school. The contributors
to the volume discuss Santos's most important concepts and apply them in different
contexts and to different topics. In the next section we briefly introduce some of
Santos's key works to give the reader a better sense of the trajectory and breadth of
the geographer's thought.

Milton Santos, in front of the Department of Geography of the University of São Paulo, 1994.
Photo by Jorge Maruta, from the collection of the magazine *Jornal da USP*

1.3   A Miltonian Epistemology

Milton Santos reinterpreted 'traditional' geographic concepts in his own way.
Concepts such as space, place, region, landscape and technique acquired completely renewed meanings in Santos's formulations. The notion of technique, in
particular, is a central organizing principle of his theories; Santos believed geography should be 'the philosophy of the techniques' (Santos 1960, 1989, 1996a).
Santos's understanding of technique is elaborated throughout this volume, especially in Chap. 11.

More than reinterpreting traditional concepts, Santos created his own vocabulary, which includes the terms tropical flexibility, used territory, counter-rationality,
fixes and flows, psychosphere and technosphere, horizontalities and verticalities,
technical unicity, the convergence of moments, the enlargement of contexts, the
knowability of the planet, contemporary acceleration, opaque and luminous spaces,

slow people, dynamic inertia, rugosity, sociospatial formation, hegemonic and hegemonized actors, territory as rule and ruled territories, and technical-scientific-informational milieu, among others. Many of these key concepts are explored by the different contributors to this volume.

In this section we present a guide of Santos's works for English-speaking scholars interested in his method, but must acknowledge that it is very challenging for non-Portuguese readers to wade through his publications. The vast majority of Santos's works are still only available in Portuguese. This is a situation that this book attempts to help rectify. There is, however, already a considerable portion of his work that has been translated and interpreted into other languages, particularly Spanish and French, and a few shorter texts that can be found in English. Engaging with these translated works contributes to a deeper understanding of his most important project, *A Natureza do Espaço: Razão e Emoção* (*The Nature of Space: Reason and Emotion*) (Santos 1996a) first published in 1996 in Portuguese, but also available in Spanish and French. At the time of writing there is still no translation of *The Nature of Space* into English, but we hope that our book motivates proficient Portuguese and English readers/writers who might be willing to take on such a responsibility in the future. *The Nature of Space*—for which Santos received the prestigious Brazilian Jabuti Award as the best book of the year in social sciences— is a compilation of Santos's most important concepts and ideas. It is therefore, a very dense and complex read. Thus, before tackling such a difficult book, we suggest that readers become acquainted with some of Santos's more straightforward and accessible works, including those presented in this volume.

Milton Santos, 1994. Photo by Jorge Maruta, from the collection of the magazine *Jornal da USP* (University of São Paulo)

An accessible introduction to Milton Santos's thought is through the documentary *Encounter with Milton Santos: Or the Global World as Seen from This Side (Encontro com Milton Santos: Ou o Mundo Global Visto do Lado de Cá) (2006)*, produced by Brazilian filmmaker Silvio Tendler. The film is in Portuguese but versions subtitled in English are also available. Tendler's documentary, in a style comparable to Michael Moore's, is based on one of the most important and most accessible of Santos's books, *Toward an Other Globalization: From Single Thought to Universal Conscience (Por uma outra Globalização: do Pensamento Único à Consciência Universal)*. This work was originally available only in Portuguese but was recently translated into English by Lucas Melgaço and Tim Clarke in a volume of this series also published by Springer (PAHSEP-12).

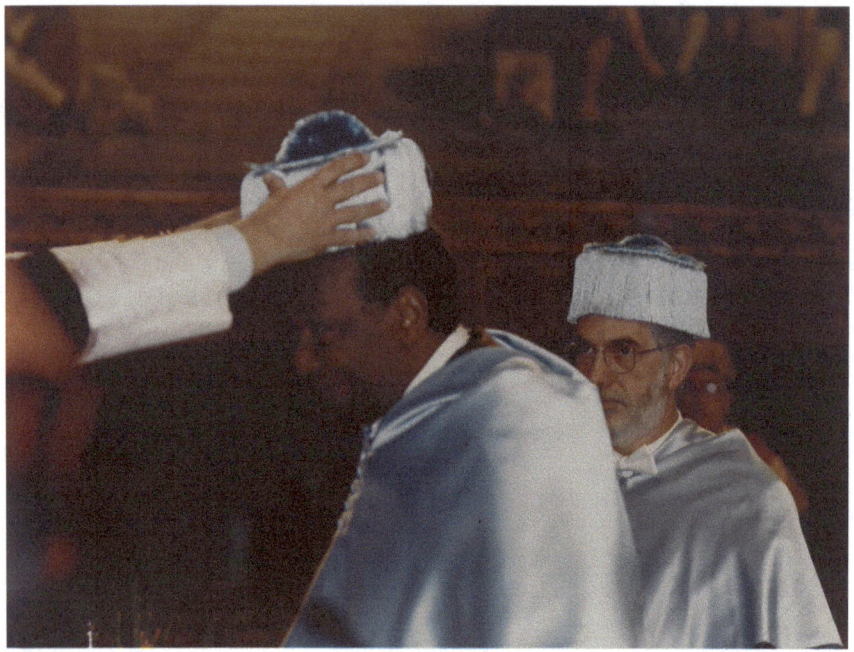

Milton Santos receiving an honorary doctorate degree from the University of Barcelona, 1996. This photograph was taken from the personal photo collection of the author and his wife, Marie-Hélène Santos, who also granted permission for publication. This and other photos can be found at: https://www.miltonsantos.com.br/

For English-reading audiences there are also a few of Santos's articles and a book that were either directly published in English or translated while the author was still alive. Most of this material, however, is from an early moment in Santos's career, mainly from the second phase described above. As already indicated, for instance, Santos's *The Shared Space: The Two Circuits of the Urban Economy in Underdeveloped Countries* (translated by Chris Gerry) (Santos 1979) became available in English in 1979 after being originally published in French in 1975 and later in Portuguese in 1978. Many in geography consider *The Shared Space* a classic. In a 1982 book review, T.G. McGee states:

> Geography is not a subject noted for its sense of adventure. Geographers tend to carry out research in well-established niches which earn the plaudits of their gurus, chairmen or funders. Occasionally, however, a geographer breaks away from these acceptable research grooves and ventures into areas at the cutting edge of the discipline. Milton Santos is such a geographer [...] Santos's book is to be commended as a pioneer investigation. It is an important milestone in the interpretation of the third world urbanization. (McGee 1982: 146)

As indicated by McGee, readers of the *The Shared Space* will find in this work Santos's efforts to create an original theoretical framework for understanding cities of the so-called 'underdeveloped' world. It is in this book that Santos introduces the concepts of upper circuit and lower circuit of economic activity. According to Santos, what differentiates the two circuits are technology and forms of organization: the upper circuit uses capital-intensive technology while the lower circuit is based on labour-intensive activity (Santos 1975). This book—and the theory presented therein —would be later partially cast aside by the author, who did not include the theory of circuits in his major book *A Natureza do Espaço*. However, Santos himself and, more recently, geographers such as María Laura Silveira and her team (Silveira 2005; Regitz Montenegro 2012), would subsequently revisit the concepts and use them to understand aspects of the contemporary socio-historical conjuncture.

The few articles Santos published directly in English are from the period in which he worked abroad. He wrote five articles for *Antipode* (Santos 1974, 1977a–c, 1980). Two articles from this list are particularly relevant for understanding the novelty of Santos's thought: 'Spatial dialectics: The two circuits of the urban economy in underdeveloped countries', based on his main ideas from *The Shared Space;* and the article 'The Devil's totality: How geographic forms diffuse capital and change social structures', a dense essay in which Santos discusses the perversities of planning in the underdeveloped world. In this latter article he also presents concepts that would reappear in other texts, such as totality, and proposes his internal categories of geography, such as form, function, structure and process. The following excerpt may give readers a taste of Santos's thoughtful and dense style:

> The study of totality impels a choice of analytical categories which must reflect the real motion of totality. We must take into account, in addition to the categories of time and scale that function externally, the internal categories of structure, function and form. The notion of process permeates all these categories. However, the process is itself nothing but a fleeting vector whose life is ephemeral; it is a brief moment, the fraction of time necessary for the structure to be realized; that is, to be geographized or better, spacialized, through a function - that is, through a more or less lasting activity, and by its indispensable union to a form. The form usually outlasts its specific function. A process ends when a fraction of the structure has been objectified in a particular form with a particular function. Then, a new process begins. (Santos 1980: 45)

English readers could also consult the English language entry about Santos in *The International Encyclopedia of Human Geography* (Souza 2009). This helpful piece is written by Maria Adélia de Souza who is a longstanding interlocutor of Santos's in Brazil and his former colleague at the University of São Paulo. The article 'Security and surveillance in times of globalization: An Appraisal of Milton Santos' theory' (Melgaço 2013) may also be of use to Anglo readers. Finally, a translation by Lucas Melgaço and Tim Clarke of the manifesto that Milton Santos and his students published in 2000 (one of the last texts he published before passing away) is also being made available by the journal *Antipode*.

In addition to *The Nature of Space*, there are a handful of books from the last phase of Santos's career that usefully explicate his thought but that have yet to be translated into English. The 1980s was one of the most productive decades of his career during which Santos published many books, three of which are worth mentioning here. *Pensando o Espaço do Homem* (1982) and *O Espaço do Cidadão* (1987) are two of Santos's most digestible texts. Because they are quite accessible they serve as useful introductions to Santos's theories. In this decade Santos also wrote *Espaço e Método* (1985; also available in Spanish and French) which, together with another book from the 1990s, *Técnica, Espaço, Tempo: Globalização e Meio Técnico-Scientífico-Informacional*, can be understood as a preparation for his culminating book *A Natureza do Espaço*, published a decade later.

Milton Santos receiving the Vautrin Lud International Prize in Geography, Saint-Dié-des-Voges, France, 1994. This photograph was taken from the personal photo collection of the author and his wife, Marie-Hélène Santos, who also granted permission for publication. This and other photos can be found at: https://www.miltonsantos.com.br/

In the last years of his life Santos published two more books, the aforementioned *Por uma Outra Globalização*, and, in partnership with María Laura Silveira, *O Brasil: Território e Sociedade no Início do Século XXI*. The latter is one of the least theoretical of Santos's books as his purpose was to apply his concepts and theories to the analysis of concrete cases in Brazilian territory.

Finally, during Santos's life (Souza 1996) and especially after his death many edited volumes, special issues, doctoral dissertations (Grimm 2011; de Ferreira 2000; Costa 2013), biographies and articles have been published about Santos. Most of these are in Portuguese with some exceptions (e.g. Zusman 2002; Scripta 2002; Levy 2007).

1.4 Crossing Linguistic Divides

This book is a project in interpretation and translation. Committing to translate ideas, concepts, theories and knowledges of Milton Santos and the scholars he inspired is inherently political. Alluding to the violences of interpretation, Paulo Bezerra in his introduction to his translation of Dostoevsky wrote, 'Every translation is the possible translation' (Bezerra 2007: 8). We recognize this tension, but also believe that the 'work of translation…seeks to turn incommensurability into difference, a difference enabling mutual intelligibility among the different projects of social emancipation' (De Sousa Santos et al. 2007: xl). In other words, translation offers the opportunity to dialogue among different experiences, which is crucial to both cognitive and social justice (De Sousa Santos et al. 2007). Our job as interpreters of Santos's ideas is to work toward this social justice while being constantly reflexive of the violences our translations may commit.

Translation and interpretation are necessarily difficult and this book has exposed to us the fraught pragmatics of doing this work. Our project is permeated by numerous lines of political economic and linguistic oppression. The contributors are, for the most part, from Brazil and do not speak or write in English as a function of their profession. As a result, some contributors paid interpreters to translate the initial versions of their chapters. Others wrote in English—itself a tall older—and the editors had to spend considerable time interpreting and shaping the language. Projects that reverse the colonial-like flows of unilateral translation (from Portuguese to English, for instance, as opposed to the much more common English to Portuguese) thus require time and money of those who often have these resources in short supply.

The work of translation and interpretation involves not simply finding the correct word or sentence structure, but also negotiating central geographic concepts and styles of writing. It quickly became apparent to us that what is taken as common sense in one linguistic tradition of geography is not always the same in the other— the 'fundamental concepts of geography' are, of course, not so fundamental. The histories of Brazilian-Portuguese geography and the canonical curriculum in this tradition are quite different than that of Anglo geography, even when there has been conversation between the two. This has meant that Brazilian Portuguese words and concepts do not always have an equivalent in English, and even where they superficially appear to be the same—such as 'space' or 'espaço'—there are different assumptions and theoretical models undergirding them. But Brazilian writing and Anglo writing are also very different. Put rather crudely, English sentence structures and conventions tend to be more utilitarian, whereas Brazilian writing—and Brazilian-Portuguese in general—is more poetic. The translations are thus not only an interpretation of key concepts, but also involve very politicized decisions as to what forms of language—such as words and sentence structures—will be received as rigorous by an Anglo academic tradition.

Temporality also shapes translation: historical shifts in linguistic conventions require political decisions be made about word choice. Santos was prolific from the 1950s until his death in the 2000s. Some of his earlier language thus appears dated to both Brazilian-Portuguese and Anglo audiences. We struggled, for instance, over whether to change Santos's use of '*homem*' (man) to 'humanity' or 'people'. We are politically committed to non-gendered language, but also believe in translating texts as true as possible to their spatio-temporal context, not wanting to over-translate them, as it were. Neither option feels entirely correct, yet we finally decided to interpret '*homem*' as 'people' or 'humanity' to preserve the gender-neutrality implied in Santos's theorizations. This struggle revealed to us the inherently 'violent' nature of translation that can never be perfectly resolved and requires considerable thought.

1.5   The Book

This book is a collection of translations and interpretations made by an array of scholars who engage with Santos in divergent ways. The first chapters are focused on the intricacies of the geographer's thought; they translate key parts of his text and explain the transformation of his ideas. Later chapters also help explain key elements of Santos's theories, but focus on applying them to particular contexts and situations often in Brazil. The chapters thus serve as an introduction to Santos's

Milton Santos at the opening of the conference in his honor, *Um Mundo do Cidadão, um Cidadão do Mundo*, at the Univesity of São Paulo, 1996. Photo by Ana Pereira

thought and also to realities of globalization in Brazil, both of which are, of course, inherently related. Below we offer a brief introduction to the chapters and layout of the book.

Among the long list of terms, notions and concepts created by Santos, the concept of *used territory* may be the most provocative. Through this concept Santos invites geographers 'to consider geographic space not as synonymous with the territory, but as *used territory*; the *used territory* is both the result of the historical process and the material and social basis of new human actions' (Santos et al. 2000: 104, emphasis in original). Unfortunately Santos passed away before fully explicating this idea. We are left with a few of his inspiring insights to be explored, developed and criticized. When deploying the concept of *used territory*, Miltonian scholars normally refer to the manifesto written by Santos and his students and from which we quote above. However, the first time Santos used the term was in 1994 in the text 'The Return of the Territory', the annotated translation of which is available for the first time to an English-speaking audience in the next chapter (Chap. 2). In this text, Santos revises and reinterprets the traditional concept of territory.[7] He argues that geographers should not study territory itself but rather what he calls the *uses* of the territory. According to the author, this choice would avoid the 'risk of alienation' and 'the risk of renouncing the future' (p. 26). The idea of a *return—* indicated in the title—refers to the fact that in spite of the growing transationalization of spaces through networks, the *used*, inhabited territories (or the banal spaces, in the words of François Perroux) create new synergies 'that challenge the world to a rematch' (p. 26). Santos develops this idea through the dialectics of what he calls *verticalities* and *horizontalities*; these concepts are central to his theoretical reflections on the global and the local.

Sarita Albagli (Chap. 3) discusses *networked space* in what the Brazilian geographer called the contemporary period: the *technical-scientific-informational milieu*. Santos conceptualizes space as a hybrid and in this particular period information technologies have become central to geographical landscapes. He explains, 'information nowadays plays a role analogous to the one played in the past by energy' (Santos 1996a: 132). Albagli offers a clear and succinct overview of Santos's approach to networked space. She pays particular attention to the relationships engendered through new *horizontalities* and *verticalities*. She also sheds light on the contributions of the geographer's thinking to the current conjuncture of global unrest: while Santos was chiefly concerned with the globalizing economic logics that institute oppressive forms of networks and telecommunications, Albagli extends his thinking by pointing to the ways that new social movements and contentious activity epitomized by mass street demonstrations are being organized also in networked, but bottom-up, fashion.

---

[7]Moraes (2013) dedicated an entire book to analyzing how Milton Santos used the concept of territory in the different phases of his career. Unfortunately, as is the case with most of Santos's scholarship, the text is thus far only available in Portuguese.

Aurélien Reys (Chap. 4), writing from France, traces the evolution of Santos's thought during the geographer's exile from Brazil. He describes how the French academy influenced Santos's theories, but also how French models were inadequate to explain the Brazilian's experiences of urbanization in his home country. It was in France that Santos began explicitly theorizing from the point of view of the Third World by conversing with, and often criticizing, dependency theory, world systems theory, notions of underdevelopment and centre-periphery relations. He also began to focus specifically on the experiences of cities in the South: how urbanization dynamics in these cities are shaped by large-scale population growth and how the city must be understood as whole system comprised of *upper* and *lower economic circuits*. These ideas are most clearly elaborated in Santos's book *The Shared Space*.

Other contributors place Milton Santos in conversation with scholars better known outside the Brazilian context. Eliza Pinto de Almeida (Chap. 5) explores useful articulations between Santos and Michel Foucault to discuss the transformation of the *technosphere* and *psychosphere* of medical care from the 18th century to the contemporary period. She explores how the space of the hospital has been transformed through shifts in the *systems of objects* and *systems of action* and the *fixed* and *flowing elements* that comprise it. Central to her argument is Santos's notion of space as an 'indissoluble, solidary and contradictory set of systems of objects and systems of actions that cannot be considered in isolation, but as the unique frame in which history unfolds' (Santos 1996a: 51). Following the historical transitions laid out both by Santos and Foucault, Almeida argues that the psychosphere of hospitals changed from fear to therapeutic care, engendered by shifts in the technosphere—specifically technological advancements in, *inter alia*, surgery, sanitation and pharmaceuticals.

Contributors to the volume also use Santos's theories to explore contemporary issues in Brazil, from the expansion of the agricultural frontier to environmental crises and new modes of nature conservation. In their chapter Samuel Frederico and Marina Castro de Almeida (Chap. 6) discuss Santos's analytic the *political economy of territory* and use one of its central tenets—the dialectical relationship between *centripetal* and *centrifugal forces*—to understand the Brazilian state's territorial organization of agribusiness in the 2000s. The authors argue that productive re-structuring and reorganization of space have been central to the large-scale development of modern agriculture. They explore, specifically, how centripetal forces are consolidating the normative, financial and informational role of São Paulo, while centrifugal forces spatially disperse many sub-spaces of monocultural production across the country. They argue that these socio-spatial dynamics have had a polarizing and unequal effect on the development of various regions in Brazil. This is, in part, because the reorganization of territory is often tied to circuits of capital that privilege global actors, thus re-drawing the territory of Brazil in the interests of these hegemons and effecting an *alienation of space*.

Milton Santos being awarded the title of Professor Emeritus of the University of São Paulo, 1997. Photo by Oswaldo José dos Santos, from the collection of the magazine *Jornal da USP*

Luís Angelo dos S. Aracri (Chap. 7) draws on Santos's works *The Nature of Space* and *Planning Underdevelopment and Poverty* to discuss two key phases of Brazilian territorial planning. In the first period, spanning the 1930s to the 1980s, the national government had an ostensible monopoly on industrial production. It centralized planning and focused specifically on integrating the national territory. From the 1990s onward the state's focus shifted as a result of new global dynamics: the national government used territorial planning to try to insert different production sectors into an increasingly globalized market and, reacting to the global market's competitive imperative, focused on attracting investments from both public and private sectors. Aracri draws specifically on Santos's notions of *fluidity imperative*, *technical networks* and *war of places* to flesh out his arguments. Aracri ends on a hopeful note by presenting an ethical imperative to imagine a non-capitalist mode of territorial planning.

Santos's understanding of space as a *hybrid* is explored in Fabrício Gallo's chapter (Chap. 8) about Brazilian federalism. Gallo utilizes Santos's notion of *used territory*, which refers to the ways geographical space is shaped by both *objects* and *actions* and by the co-constitution of *materialities* and *immaterialities*. Gallo presents the example of the Brazilian federation and explores how territory is differentially organized through federalism—how territory is differentially *used*, in other words—and the implications that this has for the unequal distribution of limited resources. He argues that the use of territory through federalism re-entrenches socio-spatial inequalities. In this chapter, Gallo explains several of Santos's key concepts including the *event, ruled territory* and *territory as rule*.

Continuing a common focus on agribusiness, Júlia Adão Bernardes (Chap. 9) uses Milton Santos's theorizations to explore modern agricultural development in Brazil. She connects the movement of the agricultural frontier—characterized largely by monoproduction—to global capital markets in the current *technical-scientific-informational milieu*. Here, science, technology and information have been central to the expansion of capitalist frontiers across Brazilian territory. Bernardes draws on Santos's concepts of *system of objects* and *system of actions* to explore the new territorial divisions of labour and reorganizations of agricultural territory in the Brazilian states of Mato Grosso and Pará. She is also concerned with how the nation-state facilitates this development through the creation of infrastructure such as the BR-163 federal highway. The modern frontier of agribusiness has had devastating impacts on the environment and has caused high unemployment rates among local populations alongside the expropriation of their land.

Roberta Carvalho Arruzzo (Chap. 10) is also interested in the expansion of the agricultural frontier but from a different locus of enunciation. In her chapter, Arruzzo discusses how agricultural production and environmental preservation efforts encroach on indigenous peoples' territory across Brazil. She draws on Santos's notion of *luminous spaces*—as those spaces of fluidity and velocity imposed by global capitalism—and his notion of *opaque spaces*—the spaces of *slow people* where counter/non-hegemonic rationalities exist. Indigenous peoples' lands in Brazil are sites of intense negotiation in which these different spaces and temporalities come into conflict. However, slow people, according to Santos, are those who offer hope for a different future, as they engage with alternative ways of being that can subvert the logics of global capitalism. They are potentially revolutionary subjects. Indigenous peoples, to Arruzzo, offer this hope.

Santos has always been very critical of a 'naïve environmentalism'. In a moment when it seemed as though almost everyone in Brazil and around the world was talking about the environmental crisis Santos was a dissenting, critical voice. He has been worried that a utilitarian geography concerned with a narrow understanding of the environment could result in a fragmented geography in which subdisciplines 'establish themselves as autonomous, when in reality their secondary role only qualifies them as operational branches of a more complex and unitary geography' (Santos et al. 2000: 106) He believes that 'both the market and politics sometimes inspire solutions' (Santos et al. 2000: 106) of naïve environmentalism. These solutions for environmental crises are realized through 'geographies of tourism, the environment, culture, geoinformatics, or of the suggestions of the so-called regional planning…[that] address only one or a few activities at any given time' (Santos et al. 2000: 106) resulting from a fragmented notion of nature. Santos does indeed look at issues of nature, but through the lens of the *technique*.

These ideas are explored in the final two chapters of the volume. Environmental preservation and sustainable development in Brazil have taken the form of conservation units (CUs) in the current *technical-scientific-informational* period. Maria Tereza Duarte Paes and Claudia Levy (Chap. 11) discuss CUs as new systems of nature regulated by specialized technical knowledge and logics such as private property, extraction and protection. These logics are *vertically* imposed, upend traditional orders and disrupt *horizontal* relationships between contiguous locations. Paes and Levy draw on Santos's concept of *technique*, the 'ensemble of instrumental and social means by which humanity realizes life, produces, and creates space' (Santos 1996a: 25), to focus on the hybrid and historically constituted condition of conservation spatializations such as those of CUs.

Santos believes that we need a nuanced and geographical understanding of Nature, not environment, to analyze space and territory. While the Brazilian geographer does not often explicitly discuss environmental crisis, his theoretical insights across disciplinary boundaries—philosophy, geography and science—are useful for understanding climate change in the contemporary period. Francisco Toro (Chap. 12) explores how the work of Santos contributes to a new ethics for responsible engagements with Nature. He does so by pursuing four major dimensions of Santos's thought: first, ontological, or the necessary historicization of Nature; second, epistemic, or the problematic assumptions of objectivity in the scientific method; third, technical, or the inherently spatial nature of technical objects and activity; and fourth, ethical, or how a spectacularized nature can compound environmental crisis. Toro's chapter helps readers understand the cross-disciplinary nature of Santos's thought through the concrete and pressing issue of climate change.

These chapters offer a small sample of the vast possibilities available to apply Santos's concepts and method. The Miltonian school uses his ideas to explain a broad range of subjects from urban violence, to finance, to music, to graffiti, to transportation. However, most of this scholarship is still trapped in language and cultural barriers that prevent English-speaking scholars from engaging with his work. We do not celebrate Santos as a god, or Miltonian thought as a dogma; his work is to be used, applied, improved and criticized. Before this happens, however, it is necessary to first read and understand his work. This book offers a step toward a serious engagement with Miltonian thought in English-language geography.

Borrowing from Santos's vocabulary we could say that the 'hegemonized' in academia, the counter-rationality of the outsiders, offer new gazes and new forms of engaging the world. Santos was critical of many exploitative processes but he was, at heart, an optimist. He believed in the *período popular da história,* in the protagonism of the poor, in globalization as possibility. But this revolutionary form of globalization must make peripherality central to a new normative ethic for living in solidarity and non-exploitative relationships with one another. Thinking of and with peripherality is indeed the *centrality* (Santos 2000a) of global social justice.

References

Bezerra, P. (2007). Nas sendas de crime e castigo. In F. Dostoiévski, *Crime e Castigo* (pp. 7–13), São Paulo: Ed. 34.

Comaroff, J., & Comaroff, J. (2012). *Theory from the South: Or, how Euro-America is evolving toward Africa*. Boulder, CO: Paradigm Publishers.

Costa, P. E. F. (2013). *O 'jovem Milton Santos': Personagem do protótipo metodológico: Revelar [matrizes clássicas originárias] para Definir [vanguarda, universalidade e viés geográfico]*. (Ph.D. Dissertation). Rio Claro: UNESP.

De Sousa Santos, B., Nunes, J. A., & Meneses M. P. (2007). Introduction: Opening up the canon of knowledge and recognition of difference. In B. de Sousa Santos (Ed.), *Another knowledge is possible: Beyond northern epistemologies* (pp. xvix–1). New York: Verso

De Araújo Ferreira, A. M. (2000). *Para um vocabulário fundamental da obra de Milton Santos: Com equivalência em francês*. (Ph.D. dissertation). São Paulo: University of São Paulo.

Festival International de Géographie. (2016). *Tous les lauréats du prix Vautrin-Lud*. Retrieved August 17, 2016, from http://www.fig.saint-die-des-vosges.fr/geographie/prix-vautrin-lud.

Grimm, F. (2011). *Trajetória epistemológica de Milton Santos*. (Ph.D. dissertation). São Paulo: University of São Paulo. Retrieved August 25, 2015, from http://www.teses.usp.br/teses/disponiveis/8/8136/tde-26062012-143800/pt-br.php.

Lawson, V. (2007). *Making development geography*. London: Hodder Arnold.

Levy, J. (Ed.). (2007). *Milton Santos, philosophe du mondial, citoyen du local*. Paris: Presses Polytechniques et Universitaires Romandes.

McFarlane, C. (2008). Urban shadows: Materiality, the 'Southern city' and urban theory. *Geography Compass, 2*(2), 340–358.

Melgaço, L. (2013). Security and surveillance in times of globalization: An appraisal of Milton Santos' theory. *International Journal of E-Planning Research, 2*(4), 1–12.

Mignolo, W. (2000). *Local histories, global designs: Coloniality, subaltern knowledges, and border thinking*. Princeton: Princeton University Press.

Mohanty, C. (2002). Under western eyes revisited: Feminist solidarity through anticapitalist struggles. *Signs: Journal of Women in Culture and Society, 28*(2), 499–535.

Moraes, A. C. R. (2013). *Território na geografia de Milton Santos. São Paulo:* Annablume.

Parnell, S., & Oldfield, S. (2014). From the South. In S. Parnell and S. Oldfield (Eds.), *The Routledge handbook on cities of the Global South*. New York: Routledge.

Regitz Montenegro, M. (2012). A teoria dos circuitos da economia urbana de Milton Santos: De seu surgimento à sua atualização. *Revista Geográfica Venezolana, 53*(1), 147–164.

Radcliffe, S. A. (2005). Development and geography: Towards a postcolonial development geography? *Progress in Human Geography, 29*, 291–298.

Rao, V. (2006). Slum as theory: The South/Asian city and globalization. *International Journal of Urban and Regional Research, 30*(1), 225–232.

Robinson, J. (2006). *Ordinary cities: Between modernity and development*. New York: USA.

Roy, A. (2010). *Poverty capital: Microfinance and the making of development*. New York: Routledge.

Roy, A. (2011). Slumdog cities: Rethinking subaltern urbanism. *International Journal of Urban and Regional Research, 35*(2), 223–238.

Roy, A. (2014). Worlding the South: Toward a postcolonial urban theory. In S. Parnell & S. Oldfield (Eds.), *The Routledge handbook on cities of the Global South*. New York: Routledge.

Roy, A. (2015). Introduction: The aporias of poverty. In A. Roy & E. S. Crane (Eds.), *Territories of poverty: Rethinking North and South* (pp. 1–35). Athens: University of Georgia Press.

Roy, A. (2016). Who's afraid of postcolonial theory? *International Journal of Urban and Regional Research, 40*(1), 200–209.

Santos, M. (1948). *O povoamento da Bahia: Suas causas econômicas*. Salvador: Imprensa Oficial da Bahia.

Santos, M. (1960). *Marianne em Preto e Branco*. Salvador: Livraria Progresso Editora.

Santos, M. (1965). *A cidade nos países subdesenvolvidos*. Rio de Janeiro: Civilização Brasileira.

Santos, M. (1974). Geography, marxism and underdevelopment. *Antipode, 6*(3), 1–9.

Santos, M. (1975). *L'espace partagé: Les deux circuits de l'economie urbaine des pays sous-developpes*. Paris: M.-Th. Génin, Librairies Techniques.

Santos, M. (1977a). Society and space: Social formation as theory and method. *Antipode, 9*(1), 3–13.

Santos, M. (1977b). Spatial dialectics: The two circuits of urban economy in underdeveloped countries. *Antipode, 9*(3), 49–60.

Santos, M. (1977c). Planning underdevelopment. *Antipode, 9*(3), 86–98.

Santos, M. (1979). *The shared space: The two circuits of the urban economy in underdeveloped countries*. London and New York, Methuen.

Santos, M. (1980). The devil's totality. *Antipode, 12*(3), 41–46.

Santos, M. (1989). O espaço geográfico como categoria filosófica. *Terra Livre, 5*, 9–20.

Santos, M. (1994). O Retorno do território. In M. Santos, M. A. de Souza & M. L. Silveira (Eds.), *Território: Globalização e fragmentação* (pp. 15–20). São Paulo: Hucitec and ANPUR.

Santos, M. (1996a). *A Natureza do espaço. Técnica e tempo. Razão e emoção*. São Paulo: Hucitec.

Santos, M. (1996b). As cidadanias mutiladas. In J. Lerner (Ed.), *O Preconceito* (pp. 133–144). São Paulo: Imprensa Oficial do Estado.

Santos, M. (2000a). *Por uma outra globalização: Do pensamento único à consciência universal*. Rio de Janeiro: Record. For the English version of this text refer to: Santos, M. (2017). *Toward an other globalization: From the single thought to universal conscience*. Translated by L. Melgaço and T. Clarke. Cham: Springer.

Santos, M. (2000b, May 7). Ser negro no Brasil hoje, *Folha de São Paulo*. São Paulo: Caderno Mais.

Santos, M., et al. (2000). O Papel ativo da geografia: Um manifesto. *Revista Território, 5*(9), 103–109.

Santos, M. (2001). Curriculum Vitae. *Milton Santos* [website]. Retrieved August 18, 2016, from http://www.miltonsantos.com.br/site/miltonsantos_curriculum.pdf.

Santos, M. (2004 [1978]). *Por uma geografia nova*. São Paulo: Edusp.

Santos, M., & Silveira, M. L. (2001). *O Brasil: Território e sociedade no início do século XXI*. Rio de Janeiro, Brazil: Record.

Scripta N. (2002). El ciudadano, la globalización y la geografía. *Homenaje a Milton Santos, 6* (124). Retrieved August 18, 2016, from http://www.ub.edu/geocrit/sn/sn-124.htm.

Silva, M. A. (2002). Milton Santos: a trajetória de um mestre. *Scripta Nova*. 2(124). Retrieved August 18, 2016, from http://www.ub.edu/geocrit/sn/sn 124f.htm.

Silva, F. S., & Silva, M. A. (2004). Uma leitura de Milton Santos (1948–1964). *Geosul, 19*(37), 157–189.

Silveira, M. L. (2005). Modernização contemporânea e nova constituição dos circuitos da economia urbana. *GeoUSP, 19*(2), 246–262.

Souza, M. A. (Ed.). (1996). *O Mundo do Cidadão. Um cidadão do Mundo*. São Paulo: Hucitec.

Souza M. A. (2009). Santos, M. In R. Kitchin & N. Thrift (Eds.), *International encyclopedia of human geography* (pp. 11–14). Oxford: Elsevier.

Sundberg, J. (2003). Masculinist epistemologies and the politics of fieldwork in Latin Americanist geography. *Professional Geographer, 55*(2), 180–190.

Tendler, S. (2006). *Encontro com Milton Santos ou o mundo global visto do lado de cá*. Rio de Janeiro: Caliban Produções, DVD (89 min), son., color.

Tiercelin dos Santos, M.-H., & Levy, J. (2011). Biografia. *Milton Santos* [website]. Retrieved August 18, 2016, from http://miltonsantos.com.br/site/biografia/.

Vainer, C. (2014). Disseminating 'best practice'? The coloniality of urban knowledge and city models. In S. Parnell & S. Oldfield (Eds.), *The Routledge handbook on cities of the Global South*. New York: Routledge

Vasconcelos, P. A. (2001). Milton Santos, geógrafo e cidadão do mundo (1926–2001). *Afro-Ásia*, 25–26, 369–405.

Wainwright, J. (2008). *Decolonizing development: Colonial power and the maya*. Oxford: Blackwell.

Yázigi, E. (1996). Milton Santos e a Criatividade, In M. A. Souza (Ed.), *O mundo do cidadão. Um cidadão do mundo* (pp. 396–445). São Paulo: Hucitec.

Zusman, P. (2002). Milton Santos. Su legado teórico y existencial (1926–2001). *Documents d'Anàlisi Geogràfica, 40*, 205–219.

Chapter 2
The Return of the Territory

Milton Santos

Translated by Lucas Melgaço, Tim Clarke and Carolyn Prouse.

Abstract (by the translators) This chapter is an annotated interpretation and translation of Milton Santos's key text, 'O Retorno do Território'. In this chapter, Santos proposes for the first time the provocative concept of 'used territory'. According to the geographer, used territory should be understood as both the result of historical processes, and the material and social basis of human actions. The idea of a return, indicated in the title, refers to the fact that in spite of the growing transnationalization of spaces through networks, the inhabited and used territories (or the banal spaces) create new synergies that challenge the world to a rematch. The text analyzes the relationships between the global and the local by invoking concepts such as verticalities and horizontalities, and the three forms of what Santos calls 'happening': homologous, complementary and hierarchical.

We live with a notion of territory that is inherited from an incomplete *modernity*[1,2] and from its legacy of pure concepts–concepts that travelled through the centuries largely untouched. It is the use of the territory, not the territory itself, that should be the object of social analysis. The territory is an impure form, a hybrid, a notion that, for this reason, requires constant historical revision. What is permanent about the

Lucas Melgaço, Assistant Professor, Department of Criminology, Vrije Universiteit Brussel (VUB), Belgium; Email: lucas.melgaco@vub.ac.be.

Tim Clarke, Ph.D. candidate, Department of English, University of Ottawa, Canada.

Carolyn Prouse, Ph.D. candidate, Department of Geography, University of British Columbia, Canada.

Milton Santos (Deceased)—Author for this chapter.

[1]Translators' note (TN): As a matter of emphasis, Santos capitalized a number of words in the original text in Portuguese. We opted, instead, to italicize them, as this is a more common convention in English.
[2]TN: In his book *A Metrópole Corporativa e Fragmentada: O Caso de São Paulo* (São Paulo: Nobel, 1990) Santos uses the metaphor 'incomplete modernity' to describe the metropole of São Paulo as a juxtaposition of traces of modernity (particularly due to economic opulence) and traces of backwardness inherited by the social and political structure. The notion can be understood thus as an inequal and selective modernization.

© Springer International Publishing AG 2017   25
L. Melgaço and C. Prouse (eds.), *Milton Santos: A Pioneer in Critical
Geography from the Global South*, Pioneers in Arts, Humanities,
Science, Engineering, Practice 11, DOI 10.1007/978-3-319-53826-6_2

territory is the fact that it is our constant living environment. It is therefore crucial to understand the territory in order to avoid the risk of alienation, the risk of losing the sense of individual and collective existence, the risk of renouncing the future.

In short, we move, through the centuries, from the ancient individual communion of places and the *universe* to the idea of today's global communion: the universal interdependence of places is the new reality of the territory. In this long journey the *nation-state* was a milestone, a turning point, enthroning a legal-political notion of territory. This notion is derived from the knowledge and conquest of the world, from the *modern state* and the *Enlightenment* to the era of the valorization of so-called natural resources.

Today, nature is historic … including the so-called 'environment'. Its 'local' value is relative, or in any case, relativized.

In the past, it was the *state*, after all, that defined places—evident from Colbert to Golbery,[3] two paradigmatic names with respect to the effective subordination of the *territory* to the *state*. The *territory* was the basis, the foundation of the *nation-state*, and this state was at the same time shaped by the territory. Today, while we experience a dialectic of the concrete world, we have evolved from the already antiquated notion of a *territorial state* to the postmodern one of the transnationalization of the territory.

However, just as before not everything was, shall we say, 'statized' territory, today not everything is strictly 'transnationalized'. Even in places where the vectors of globalization are more coherent and effective, the inhabited territory creates new synergies and ends up challenging the world to a rematch. The active role of the territory makes us think about the beginning of *history*, although nothing is as it was before; hence, the metaphor of the rematch, the return.

Again, we must insist on the relevance, today, of the roles of science, technology and information. If we want to deepen the process of knowledge about this aspect of the total reality, it is not enough to talk about mundialization or globalization where the territory is concerned. The territory is made up of forms, but the used territory[4]

---

[3]TN: Jean-Baptiste Colbert (1619–1683) was a French engineer, geographer, economist and politician who served as the Minister of Finances during the reign of King Louis XIV. Among his concerns as a statesman and strategist was the construction of roads and canals to propel the country's economy, as well as Brazil's competition with its neighbours and with England. Golbery do Couto e Silva (1911–1987) was a general in the army and one of the most important theoreticians and strategists behind Brazil's 1964–1985 military government. He was the creator of the Brazilian National Security Doctrine and in 1966 he published the book *Geopolítica do Brasil*, which strongly influenced the armed forces during the military dictatorship.

[4]TN: The abstract concept of 'território usado' suggested by Santos could be translated into English as either 'used territory' or 'territory in use'. In the text 'The active role of Geography: A manifesto' Santos et al. elaborate further on the definition of the concept: 'An approach that considers the idea of the *used territory* leads to the idea of *banal space*, everyone's space, the whole space. It is the space of all humanity, regardless of its differences; the space of all institutions, regardless of their strength; the space of all companies, regardless of their power' (Santos et al. 2000: 104). See Santos, M. et al. (2000). O Papel ativo da geografia: Um manifesto. *Território*, 5(9), 103–109. A translation of this text into English was published by Antipode:

is made up of objects and actions, and is a synonymn for human space, inhabited space. Even the analysis of fluidity[5]—the latter which serves a competitiveness that now governs economic relations—operates along the same lines. On the one hand, we have a virtual fluidity that is offered by objects created to facilitate this flow, which are increasingly technical objects; on the other, objects give us only a virtual fluidity, because real fluidity derives from human actions that are becoming increasingly informed, regulated, normatized.[6]

This reality allows the identification of new divisions[7] of the territory today, cuts that go beyond the old category of the region. This is a result of the new construction of space and the new functioning of the territory through what I am calling horizontalities and verticalities. The horizontalities are the domain of contiguity, those neighbouring places that meet through territorial continuity. The verticalities are formed by points distant from each other, which are connected by all social forms and processes. From this scheme, we must reclaim François Perroux's idea of banal space, which he bequeathed to geographers and which he suggested to his disciple, Jacques Boudeville, could be tested in Brazil. The idea of banal space, more than ever, must be raised in opposition to the notion that is currently gaining ground in the territorial disciplines: the network.

Networks constitute a new reality that, in some ways, justifies this vertical formulation. But besides the networks, before the networks, despite the networks, after the networks, with the networks, there is the banal space, the space of all, the entire space. The networks constitute only a fragment of the space and the space of a few.

Today, the territory may be formed by both contiguous and networked places. However, the same places that form networks also form the banal space. They are the same places, the same points, but which simultaneously contain different functionalities, perhaps divergent or even opposite ones.

(Footnote 4 continued)

[*]Santos, M. et al. (2017). The active role of geography: A manifesto. *Antipode, 49*(4). Retrieved February 1, 2017, from http://onlinelibrary.wiley.com/doi/10.1111/anti.12318/abstract.

[5]TN: Santos's concept of fluidity of the territory encompasses both transportation (of goods and people) but also that of information (communication of data, money, ideas, orders, etc.). 'One of the characteristics of the world today is the exigence of fluidity to the circulation of ideas, messages, products or money, in the interest of hegemonic actors. The contemporary fluidity is based on technical networks, that are the foundations of competitivity. Hence: the voracious search for even more fluidity, which leads to the search for even more effective techniques. Fluidity is at the same time cause, condition and result' (Santos, M. (1996). *A Natureza do Espaço: Técnica e Tempo, Razão e Emoção*. São Paulo: Hucitec. p. 218).

[6]TN: We have added the word 'regulated' here, not present in the original Portuguese, in order to ensure the reader fully understands what Santos means by 'normatized' ('*normatizado*', in the original text).

[7]TN: In fact, Santos uses the word '*recortes*' (cut or snip) which in this context can refer to both the divisons of the territory and to the different scales of analysis: a global cut, a local cut.

This simultaneous happening,[8] made possible thanks to the miracles of science, creates new solidarities: the possibility of a solidary happening, despite all forms of difference between people and between places.

In fact, this solidary happening presents itself in three forms in the current territory: a homologous happening, a complementary happening and a hierarchical happening.

The homologous happening is related to the areas of agricultural or urban production that are modernized through specialized information. These areas lead to a rationality of behaviours presided over by the same information that creates a similitude of activities, producing functional contiguities that give the contours of the area thus defined. The complementary happening is that of the relationship between city and countryside and of the relationships between cities, which is also a result of the modern needs of the production process and of geographically close exchange. Finally, the hierarchical happening is one of the results of the tendency toward the rationalization of activities; it occurs under a command, an organization that tends to be concentrated, which forces us to consider the production of this command, its direction. This rationalization of activities under a concentrated command also contributes to the production of a sense, a meaning, imprinted on people's lives and in the life of the space.

In all cases, information has a role today that is similar to the one that belonged to energy in the remote past. Previously, especially before human existence, what brought together the different parts of a territory was the energy that originated from natural processes themselves. Over the course of history it is information that gains this function, to the point that it is today the true instrument of union between the various parts of a territory.

For the homologous happening and the complementary happening, that is, in the areas of homologous production in the city and in the countryside, the current territory is marked by an everyday life lived by rules that are formulated or reformulated locally. In this case, it is the use of information that tends to generalize. The hierarchical happening, on the contrary, refers to an everyday life that is imposed from outside, led by privileged information—an information that is secret, that is power itself. In the homologous and complementary happenings, we have a dominion of forces that are locally centripetal. The hierarchical happening, on the contrary, is the domain of centrifugal forces. There is undoubtedly in the latter case a centripetalism, but it is a centripetalism of the other.

In the first and second hypotheses, forms are primary, but the techniques are also relevant here, as they incidentally produce the forms used. In the case of the hierarchical happening, norms are primary, but it is politics rather than the techniques that is relevant.

The dialectic in the territory is thus reinforced. We might even say the dialectic of the territory since, by being used, the territory is human and can thus involve a

---

[8]TN: Santos uses the abstract term 'happening' ('*acontecer*', in Portuguese) with the global meaning of how actions, things, people and places interact.

dialectic. This dialectic affirms itself through a 'local' control of the 'technical' portion of production and a remote control of both the 'technical' and political portions of production. The technical portion of production allows local or regional cities to have some control over the lands that surround them. This command is based on the technical configuration of the territory, on its technical density and also, in some way, on its functional density, which can also be called informational density. On the other hand, distant control, locally realized through the political portion of production, is accomplished by global cities and their relays in the various territories. The result is the acceleration of the process of alienation of spaces and humanity,[9] of which a component is the tremendous mobility currently available to people: that maxim of Roman law, *ubis pedis ibi patria* (where my feet are, there is my homeland), has lost or changed its meaning today. Additionally, local and international law are changing in order to recognize the right of those who were not born in a particular place to intervene in that place's political life.

It is necessary to reflect on the conflict between, on one hand, the act of producing and living (a function of the direct process of production) and, on the other, the forms of regulation related to other instances of production. So, in our time, the importance of labour is renewed, conditioned by the technical configuration of the territory in rural and urban areas, and connected to the immediate process of production. The results of this labour are an important fact in understanding the present society.

Consequently, there is an escalating conflict between local space, a space lived by all neighbours, and global space, inhabited by a rationalizing process and with an ideological content of distant origin. This process and content arrive at each place with objects and rules established to serve them; hence, the interest in reclaiming the notion of banal space, that is, the territory of everyone, which is often contained in the limits of the work of everyone. The arrival of these objects and rules also accounts for the interest in counterposing the notion of banal space to that of networks, the latter being the territory of forms and rules for the service of a few. There is an opposition, therefore, between the entirety of the territory and some of its parts, or points, which is to say, the networks. But what produces, what commands, what disciplines, what normalizes, what imposes a rationality upon networks is the *world*. This world is that of the universal market and of world governments. The International Monetary Fund, World Bank, General Agreement on Tariffs and Trade, international organizations, global universities, the foundations that foster research by giving money, are all part of an intended world government. They give grounds to a perverse globalization and to attacks made today, in practice and ideology, against the *territorial state*.

When one says *world*, one is speaking mainly about the *market*, which today, unlike before, permeates everything, including people's consciousness: we speak of

---

[9]TN: Santos often uses '*homem*' (man) as the general term for humanity. We have decided to interpret '*homem*' as the gender-neutral 'humans' or 'people' because Santos meant all of humanity, not strictly men. See Chap. 1 for a discussion of the politics of interpretation and translation.

the market of things, including nature; the market of ideas, including science and information; the political market. The political version of this perverse globalization is precisely the democracy of the market. Neoliberalism is the other arm of this perverse globalization, and both of these arms—the democracy of the market and neoliberalism—are mobilized to impede the affirmation of those ways of living whose solidarity is based on contiguity, in the solidary neighbourhood—in other words, in the shared territory. If this coexistence is exposed to an external regulation, it is combined with national and local forms of regulation. The conflict between these norms should be considered a fundamental basis of geographic analysis.

Before the weakening of the current *territorial state*, the scale of the technique and the political scale overlapped. Today these two scales are different and distant from each other. Therefore, the great contradictions of our time emerge when we take the use of the territory into account.

In the democracy of the market, the territory is the support for the networks that transport rules and norms that are utilitarian, partial, partialized, and selfish (from the point of view of hegemonic actors); hence, the proliferation of verticalities. At the same time, horizontalities, though weakened and with limited force, are required to take account of the totality of actors.

The arena of opposition between the market—which singularizes—and civil society—which generalizes—is the territory, in its various dimensions and scales.

For now, the place—no matter its dimension—is the locus of resistance of civil society. However, nothing prevents us from devising ways of extending this resistance to larger scales. In this respect, it is essential to stress the need for a systematic knowledge of reality by analyzing its fundamental aspect, the territory (the used territory, the use of the territory). Foremost, it is necessary to re-envision reality from within, in other words, to question its own constitution at this historic moment. The discourse and the metaphor, that is, the literaturization of knowledge, can and should come later.

The (transnationalized) territory reaffirms itself by way of the place and not only through the new basis of the space, in spite of the new fundamentalisms of the fragmented territory—e.g. new nationalisms and localisms.

It is nevertheless important to remember that, thanks to the miracles enabled by science, technology, and information, the very forces that create fragmentation may, in other circumstances, serve goals precisely opposite to that fragmentation.

The current trend is for places to unite vertically. Everything, everywhere, is being done in favour of this unification. International credit is made available to the poorest countries in order to allow networks to establish themselves in the service of big capital. But places can also unite horizontally, rebuilding a basic common living that is capable of creating local regulations, regional norms…

In the vertical union, the modernization vectors are entropic. They bring disorder to the regions where they settle because the order they create is for their own exclusive and selfish benefit. If they increase horizontal cohesion, it is at the service of the market, though this process tends to erode the horizontal cohesion that serves civil society as a whole.

Yet the effectiveness of this vertical union is always at play and cannot survive except at the expense of strict norms—despite the existance of neoliberalism. In the Brazilian case, these rigid norms to which we have been submitted for ten years[10] come at the sacrifice of the nation.

Meanwhile, horizontal unions can be enlarged by new forms of production and consumption. An example of this enlargement is how farmers come together to defend their interests, allowing them to move from a purely economic form of consumption, necessary to their production, to a locally defined political consumption, which also distinguishes the Brazilian regions from each other. We should bear this in mind when thinking about the creation of new horizontalities. These horizontalities will, from the foundation of the territorial society, find a way to release us from the curse of the perverse globalization in which we are living. They bring us closer to the possibility of creating an other globalization, one capable of restoring humankind to its dignity.

---

[10]TN: It is worth remembering that the text was originally published in 1994. Santos thus refers to the neoliberal governments that came after the Brazilian military dictatorship that formally ended in 1985.

Chapter 3
Technical-Scientific-Informational Milieu, Networks and Territories

Sarita Albagli

Abstract This chapter sets out to revisit the conceptual and analytic framework developed by the Brazilian geographer Milton Santos since the 1980s and more systematically from the mid-90s. At this time Santos proposed to reflect on the role of information, science and technique in the contemporary geography of socio-technical networks and their imbrications in territorial dynamics. The chapter begins by presenting the concepts of space and territory adopted by Santos, considering the complementary relations between the 'geography of flows' and the 'geography of fixed objects'. It characterizes what the geographer calls the technical-scientific-informational milieu and its increasing relationality through a new geography of networks. The chapter also contextualizes the distinction between vertical dynamics of information in networks, and horizontal dynamics of communication within the territory; the disputes between hegemonic rationalities and emerging 'irrationalities', or 'counter-rationalities'. To conclude, the chapter proposes new elements for research agendas on the contemporary scene inspired by Santos's theoretical contributions, considering the emergence of new issues and their implications for renewing the author's conceptual and theoretical repertory and frameworks. The chapter points to the fact that digital networks, as relational and interactive environments, are inseparable from power relations. On the one hand they express state and corporate control. On the other, they propitiate collective, collaborative, and social action and production.

Sarita Albagli, Senior Researcher, Brazilian Institute of Information in Science and Technology (IBICT), Brazil; Email: sarita@ibict.br. Translation from Portuguese by Maria Cristina Matos Nogueira.

© Springer International Publishing AG 2017
L. Melgaço and C. Prouse (eds.), *Milton Santos: A Pioneer in Critical Geography from the Global South*, Pioneers in Arts, Humanities, Science, Engineering, Practice 11, DOI 10.1007/978-3-319-53826-6_3

3.1 Introduction

This chapter sets out to revisit the conceptual and analytic framework developed by
Milton Santos since the 1980s and more systematically from the mid-90s, when he
labelled the configuration of geographical space a 'technical-scientific-informational
milieu'.[1] From this point onwards Santos reflected on the role of information in the
contemporary geography of socio-technical networks and their imbrications in territorial dynamics.

Santos's theoretical contributions are taken in order to come to grips with the
simultaneously cooperative and conflicting dimensions of these new spaces of
sociability. On the one hand, these environments are relational and interactive,
fostering collective and collaborative production within which a 'universal intelligence' (Santos 1997: 158) is engendered. In other words, they create new possibilities of knowledge about the planet and of communication between all places
through information. In the author's own words, production is characterized
nowadays as fundamentally 'work upon work' (Santos 1997: 203). On the other
hand, these environments are inseparable from the issue of power; they are environments within which centralized control and technical unity as well as spatial
heterogeneities are expressed.

In this chapter I revisit relationships that Santos (1997) theorized between the
'geography of flow' and the 'geography of fixed objects'; between the 'informational density' of networks and the 'communicational density' of territories;
between networks and territories as spaces of opposition between, on the one side,
verticalities and selectivities, and, on the other, horizontalities based on the totality
of actors and actions; between the moulding of hegemonic rationalities and the
emergence of 'irrationalities', 'counter-rationalities' or yet, 'other forms of
rationalities'. In sum, I seek to demonstrate the topicality and the pioneering
character of Santos's thinking regarding geographic space. I also point toward new
theoretical elements and research agendas inspired by his contributions and
reflections. In so doing, I take into account the emergence of new issues, specifically those that arise from the innovations associated with the development and
dissemination of information and communication technologies; as well as the
potentials for renewing Santos's repertoires and theoretical-conceptual frameworks
in this context.

This chapter begins by presenting the concepts of space and territory adopted by
Santos. Next, it characterizes what the author calls the technical-scientific-informational
milieu and its increasing relationality through a new geography of networks. Further, it
contextualizes the distinction between vertical dynamics of information and horizontal

---

[1]Santos had been working on this topic since the mid-1980s, but it is in his book *A Natureza do
Espaço: Técnica e Tempo; Razão e Emoção*, the first edition of which is from 1996, that he
develops this approach more thoroughly; this is followed by the book *Por uma Outra
Globalização: Do Pensamento Único à Consciência Universal*, published in 2000.

dynamics of communication in the territory. To conclude, I propose an agenda of possible reflections on the contemporary scene based on Santos's theoretical contributions.

3.2 Space, a Hybrid

Santos (1997) characterizes space as a frame for life, made up of the indissociable union of systems of objects and systems of action. It is life, it is society in movement that fills and animates space and that is capable of transforming it and providing it with dynamism. Space is, then, a mixture, a hybrid, a forever-provisional synthesis between, on the one hand, spatial shapes and territorial configurations and, on the other, content and social relationships. These shapes, configurations, contents and relationships all interact and affect one another.

Santos, then, does not think of territory in *itself* as a theoretical construct. '[Territory] only becomes a useful concept for social analysis when we consider it as it is used, from the moment when we conceive it together with those actors who use it' (Santos 2011: 22).[2] What matters, then, is territory that is used, acted upon.

The material basis of territory—comprised of its objects—is a condition, a limit and an invitation to act. It is also the starting point of our theorizations: 'Nowadays we do not achieve anything if we do not start from the objects that surround us' (Santos 1997: 257). For Santos, material objects have, on the one hand, an autonomous corporeal existence that ensures continuity in time. However, on the other, they do not have a life of their own: 'they have neither a history nor a geography…their meaning is always relative' (Santos 1997: 82). It changes throughout history: 'An object considered on its own has value as a thing, but its value as social data arises from its relational existence' (Santos 1997: 124). Actions attribute new meaning to existing objects. Objects may be 'already acted upon' and qualified by previous actions or they may be created through new actions which, in turn, make possible new forms of both object and action. Thus, systems of objects and systems of actions condition and affect each other.

At each historical period new systems of objects and new forms of action correspond to new technical systems. Technical systems refer to a given state of the techniques, that is, ways to produce energy, goods and services, and also ways to mediate relationships between people. However, historical transformations are not subsumed only under a technical rationality. 'There is always a measure of imponderability in the result, due, on the one hand, to human nature, and on the other, to the human character of the milieu' (Santos 1997: 76). This imponderability is what characterizes the *event,* a concept to which Santos attributes great importance in geographical thought. 'An event is the result of a cluster of vectors,

---

[2]All quotes from Santos were translated from Portuguese into English by Maria Cristina Matos Nogueira.

conducted by a process, giving a new function to the existing milieu' (Santos 1997: 76). Its main feature is the unpredictability of results. Santos believes that the event is not only a temporal process, but also a geographical one: it is 'only complete when it is integrated into the milieu' (Santos 1997: 76). The event can also be perceived as inter subjective praxis[3] leading to change. Santos (1997: 116) highlights the fact that 'events dilute things'—as well as our established knowledge—demanding new knowledge. It requires new objects and new ways of thinking.

3.3 Technical-Scientific-Informational Milieu

Santos indicates the existence of three major historical stages in what he considers to be the growing artificialization and instrumentalization of the geographic milieu. These are: the natural milieu stage, the technical milieu stage and the technical-scientific-informational milieu stage. The last corresponds to the current period that starts after World War II and becomes more prominent in the 1970s.

In the technical-scientific-informational milieu, the system of objects and the system of actions are thoroughly saturated with science and technology. This implies a 'scientifization and a technicization of the landscape' (Santos 1997: 191). One of this period's most important traits is the emergence of a *techno science* in which science and technique become inseparable elements as well as highly selective, that is, they often serve private interests and have exclusionary uses. The increasingly globalized market conditions the use of this techno science to a large extent.

As an ensemble of fixed objects and flows, space is, in this contemporary stage, made up of: a technosphere—the world of objects that intensively uses science and technology; and a psychosphere—the realm of ideas that is the locus for producing meaning, the sphere of intersubjective action. Space is an ensemble of hybrid objects whose corporeal existence is only made real through actions. 'The technosphere and the psychosphere are the two pillars through which the scientific-technical milieu introduces rationality, irrationality and counter-rationality into the very content of the territory' (Santos 1997: 204). To Santos, rationality means a purely instrumental perspective of use of the territory. Irrationality and counter-rationality are forms of resistance to this instrumental use and of directly or indirectly proposing alternative ways to hegemonic logics of relationships with territory.

Information technologies such as cybernetics, telecommunications and micro-electronics are central to the current informationalized technosphere. They possess a unifying character and they act in an integrated and connected way. They also have an invasive character, as they expand and impose themselves globally. These technologies extend themselves through the productive apparatus—as well as

---

[3]Santos refers to 'transindividual praxis' as discussed by Gilbert Simondon (1958) in *Du Mode d' Existence des Objets Techniques*. Paris: Aubier.

through territories—making use of and reproducing networks for their own ends. Santos argues that 'informationalization' in the technical-scientific-informational milieu expresses fundamental changes in social and economic life—here, life is increasingly becoming 'intellectualized' as a product of intellectual labour. In other words, the dominant form of production, particularly in cities, involves work upon work, or the manipulation of meanings and of people.

Today, technical objects are also informational in character because they are invested with an informational intentionality. In other words, 'the main energy fuelling their operation is also information' (Santos 1997: 190). The scientifization and technicization of space correspond then to its informationalization: both the objects that configure the materiality of space (fixed objects) as well as the actions that provide it with life and movement (flows) are rich in information. In fact, to Santos, everything we produce nowadays is, in some way, information. However, Santos (1997: 205) warns that 'information only occurs with action…Objects—even when they are constitutionally rich in information—can never be *acted upon*. They remain in a state of rest or inactivity—waiting for an actor.' Producing a technical object does not imply its immediate use, 'until the social energy includes it in the movement of life' (Santos 1997: 173).

To Santos (1997: 132), information nowadays plays a role analogous to the one played in the past by energy: that of acting as a 'real instrument for bringing together the different parts of a territory.' Thus the technical-scientific-informational milieu ensures the operation of economic, cultural and political interlinked processes, which universalizes the geographical milieu thus making it global, or, better, moulds 'spaces for globalization' (Santos 1997: 191). Remember that, according to Santos, 'the geography of flows depends on…the geography of fixed objects' (Santos 1997: 213). In this contemporary context, the intensity and importance of (material and immaterial) flows and of interactions has increased, mediated by technique.

The political dimension is also central to the new milieu. Following Santos's theory, contemporary globalization occurs through the 'confluence between new technical conditions and new political conditions' (Santos 2011: 27). Increased flows and circulations configure and prevail upon production processes, turning speed into an imperative, a political model for civilization. Speed is, above all, political rather than technical. Likewise, fluidity is a socio-technical entity of a mixed, relative and selective character. It benefits people and places unequally. Fluidity becomes an attribute that differentiates subspaces as those that are easily updated and those that tend to become outdated. Thus, fluidity keeps space 'united, but differentiated' (Santos 1997: 253). Objects and places are nowadays created to facilitate flow. 'We can also say that they 'circulate'. It is as if they [themselves] also were flow' (Santos 1997: 218). The internet of things is the best expression of this process: it creates the capacity to connect objects and to make them interactive in a sensorial and intelligent way through the World Wide Web. To Santos, fluidity is an attribute, the possibility or capacity of circulating, while flows are the circulation processes themselves.

Santos (2011) argues that the major pillars of contemporary globalized capitalism are the 'tyranny of money' and the 'tyranny of information'. He refers to the 'imperative', the 'omnipresence', the 'despotism' and even the 'violence' of information to characterize the appropriation of information techniques by states and businesses, the centralization of information by a limited number of companies, and the manipulation of information that is passed on to a considerable number of human beings. To Santos (2011: 39), 'information has two sides: one that seeks to instruct and one that seeks to persuade'. Regarding the latter, Santos refers specifically to the role of advertising in anticipating production and the role of media in the 'interested or otherwise self-interested, interpretation of facts' (Santos 2011: 41). However, he also believes that information may carry the prospect of revolution for our time: information potentially allows for the world to become known everywhere. It makes possible 'instant awareness of what happens to others' (Santos 2011: 28) and, therefore, also makes possible a growing universal awareness.

3.4   The Geography of Networks

Within the technical-scientific-informational milieu, space is increasingly moulded by relationality. The reticular space of socio-technical networks pervades and even overlays territories. As such, today there are distinct forms of relationship between networks and territories, such as: territories of networks, networks of territory, networked territories or territory as social network whose dynamism is connected to flow.

Santos assigns to networks a series of attributes in the contemporary period:

- they are not merely a technical and material reality: they are also human, 'living', social and political; they are made up of objects and actions whose materiality remains as a mere abstraction until it is activated by 'the people, messages and values that frequent it' (Santos 1997: 209);
- they constitute a phenomenon that becomes absolute, but not total: 'not everything is a network' (Santos 1997: 213); the space of flows constitutes only a subsystem made up of dots and lines;
- they are increasingly more deliberate and less spontaneous;
- they are simultaneously global and local; it is at the local scale that 'fragments of networks gain a unique and socially concrete dimension' (Santos 1997: 215);
- they are not uniform: they are heterogeneous and differentiated, one and multiple; it is the world that awards them their primary unity, but also their plurality as they take on diverse forms in space and time;
- they are active, not passive; they are stable and dynamic; their dynamism issues from the movements of society;
- they integrate and disintegrate the order of the territory and of spatial divisions, thus creating other orders and divisions;

- they encompass and reverberate different logical systems and types;
- their existence is inseparable from power; the role of actors in the control of their operation and use is diverse;
- they are hybrids of a mixed and ambiguous nature.

As Santos summarizes:

> Networks are virtual but at the same time real. As with all and any technical object the independent material reality of networks is a promise…[T]he first characteristic of networks is that they are virtual. They only become real - really effective, historically valid - when employed in the process of action. (Santos 1997: 220)

3.5 Horizontalities and Verticalities

Santos establishes a difference between the dynamics of information and of communication. He believes that informing and communicating are not equivalent and do not necessarily occur simultaneously. This distinction is related to horizontalities and verticalities in a territory. The dynamics of information are related to verticalities. Verticalities to Santos are relationships between locations that are spread out in space. The 'ensemble of locations creates a space of flows' (Santos 2011: 105)—that is, a reticular, networked space. Informational density indicates 'the degree of exteriority of the place and the fulfilment of its tendency to engage in a relationship with other places privileging sectors and actors' (Santos 1997: 205). Information, then, directly affects the vertical integration of diverse locations.

Santos associates the dynamics of communication with the horizontalities of contiguous relationships within a territory. Here, both the finality—that is, a territorial project or goal—imposed from the outside and from above, and the counter-finality engendered locally, co-exist. It is the banal space[4]—the space for living—that contains ensembles of hegemonic and non-hegemonic actors, the presence of myriad rationalities and the coexistence of diversity, cooperation and conflict. Horizontalities 'are the theatre of a conforming, but not necessarily conformist, quotidian. They are simultaneously the place for blindness and for discovery, for complacency and for rebellion' (Santos 1997: 227). Santos (1997: 253) quotes the philosopher Henri Laborit who reminds us that 'communication means etymologically to make in common'[5]; this leads to the creation or the fostering of

---

[4]Santos found inspiration in François Perroux's (1961) discussion of banal space, which was also appropriated by Manuel Castells at a later stage. Perroux labelled banal space as that which was not limited by flow or economic actors.

[5]The quote by Henri Laborit is from *L'Homme et la Ville*. Paris, Flamarion (1987: 38).

social bonds, of reciprocity and cooperation bringing about unity and diversity[6] across the territory.

To Santos, large cities are 'the great banal space'—a space of socio-spatial diversity, of a wealth of perspectives, of mixed interpretations. Cities are also places of great mobility and the possibility of encounters. Anyone can settle in large cities; this broadens both the division of labour and the paths for interaction and inter-subjective relationships occurring within them. This also allows for the construction and redesigning of values. Cities are, in effect, 'where the weak can subsist' (Santos 1997: 258).

Santos's theorizations identify, then, a desire for a new type of politics—bottom-up politics—that exists outside institutional structures and expresses itself as much through violence as 'through the desire for understanding and overcoming' (Santos 2011: 60). The present moment, ridden as it is with conflicts, constitutes fertile soil for the creation of new and more numerous possible futures. Santos saw potential in the excluded. To him, those not benefiting from material modernity—that is, the poor, migrants, minorities—exist in less modern, 'opaque' areas versus 'luminous' zones or 'spaces of exactitude' (Santos 1997). As such, they are the sources of irrationalities, or better, of counter-rationalities, of other rationalities or parallel rationalities. Because they escape hegemonic rationalities, which leave little room for spontaneity and variety, the excluded are the source of creativity and of future possibilities. Santos believes that it is the poor in the city who can inspire new debates and who can occupy open spaces and spaces of creativity. 'Fundamental destitution'—in the words of Sartre (1960)—or scarcity—as a permanent feature of existence fed by the mismatch between desired but unfulfilled consumption—generates 'creative discomfort': 'the wealth of 'have-nots' is the readiness of senses' (Santos 2011: 130). This is because,

> The less an individual belongs - by being poor, part of a minority, a migrant... - the more easily the shock of novelty hits her and the discovery of new knowledge comes more easily to her...The more unstable and full of surprises the space is the more surprised the individual will be; and the more effective the operation of discovery. (Santos 1997: 263–264)

The poorest areas and social groups possess greater flexibility and capacity to adapt to change as well as motivation for the construction of solidarities and the pursuit of greater freedom: elements that 'however much you give them away, the more you have of them' (Santos 2011: 130).

To Santos, then, ephemera and discovery are more important than long-lasting experience and territoriality. Santos (1997: 261) believes that poor and migrants are capable of finding 'new uses and functions for objects and techniques as well as new practical articulations and new norms in both social and affective life'. Like other social segments the excluded appear to be passive when faced with technical

---

[6]Milton Santos develops here a broad reflection on the relationship between universal consciousness and individual existence within relationships of reciprocity, bringing forth otherness and communication. He recovers the idea of transindividuality—proposed by Gilbert Simondon—as relationships between human beings mediated by technique.

and informational networks and dynamics. However, they are very active within the communication sphere, and their interaction contains more content. Importantly, though, this insight of Santos's has largely been contradicted by the way counter hegemonic movements have made use of digital platforms as instruments of mobilization in their struggles. Santos on the one hand could see and foresee the uses of digital media in the surveillance and control of citizens and in labour exploitation. On the other, he could not witness the expansion of the uses of these media as forms of alternative circulation of information and social mobilization as well as of collective and non-market forms of social production and collaboration.

Even though he criticizes the rationality inherent to dominant objects and practices in the technical-scientific-informational milieu, Santos (2011) glimpses the potential of information technologies to make possible new alternative paths to change. He acknowledges that the technical informational apparatus, which he calls 'soft techniques', demands less fixed capital and greater intelligence; it is more adaptable to different milieus and cultures and spread more easily through the social body. It could possibly even lead to less economic concentration. Santos believes that 'thanks to the lightning progress of information the world becomes closer to each one of us no matter where one is'; thus information establishes proximity with others and generates 'awareness of being world and of being in the world' (Santos 2011: 172)—a universal awareness.

To achieve this integrated awareness, Santos argues that the democratization of these soft techniques is necessary: they must be removed from their subordination to great capital and placed at the service of humanity. Along these lines, he foretells: 'It is enough that two major mutations currently being engendered come into being— the technological mutation and the philosophical mutation of the human species' (Santos 2011: 174). The technological mutation is associated with the democratization of information techniques in the service of humankind. The philosophical mutation may give a new and sustainable meaning to existence of each person and also the planet.

3.6 New Frontiers

The work of Milton Santos was pioneering on different fronts and from different perspectives. It not only introduced to geographical thought topics and issues little explored in this field, but it was also taken up by other disciplinary arenas and fields. In particular, the author demonstrated the centrality of the spatial dimension for understanding contemporary transformations.

In this chapter, my purpose has been to present Santos's contributions to understanding the role of contemporary hegemonic technical systems—information and associated technologies in particular. He had much to say about the connections of current political and social transformations to spatial dynamics.

Santos primarily focused on bringing to light the perverse aspects of the relationship between the technical-scientific-informational milieu and the intensification of the globalization process. However, he also acknowledged the potentialities that were opened up by alternate rationalities and placed importance in a universal awareness guided by a new valuation of otherness. He foresaw the potential of a new bottom-up type of politics that begins with the excluded and the marginalized, outside established institutional boundaries.

Central to Santos's ideas was that territorial dynamics involve the coexistence of diversity, heterogeneity and inequality in the territory and between territories. Moreover, he showed how cooperation and conflict coexist within territorial dynamics: cooperation through the solidarities and synergies engendered in the context of life *in common* within the territory; conflict through different meanings and disputes in using and appropriating the territory.

Santos also sought to demonstrate how territorial dynamics and those of socio-technical networks affect each other while preserving their specificity. Networks are selective, they select and establish certain connections, while territory is banal space—all the space. With new information and communication technologies that form the technical basis for globalization, territories are redefined, acquiring informational and technical density. In turn, from the perspective of networks, territories are hubs of global communication within which agents for social change are concretely based.

Milton Santos could not witness the later evolution of these tendencies. They would come to shape new socio-digital formations, spaces and dynamics; they are now the new battlefields, the true laboratories and the advanced frontiers of social and political experimentation and innovation. On the one hand there is the contemporary intensification of informationalization being used by powerful forces through data collection in the era of *big data* and its new ways of capturing socially created value; the new mechanisms for surveillance and control; and the strengthening of systems for protecting intellectual property. On the other hand, social movements are using digital networks for counter-hegemonic purposes through counter-surveillance and inverse surveillance; alternative pathways for circulating information outside large corporate media; and many other fronts of antagonism and struggle (Albagli & Maciel 2010).

One of the core questions for democracy nowadays concerns the socialization of knowledge, information and culture, on the one hand, and the appropriation and privatization of these flows, on the other (Albagli & Maciel 2013). This tension is the source of many disputes and antagonisms. Democratic concern with these processes brings to light a new agenda on rights; it mobilizes new actors and new political and social protagonisms. Milton Santos's thinking provides a valuable treatise on the possible courses of action within this contemporary context of change.

References

Albagli, S., & Maciel, M. L. (2013). Informação, conhecimento e democracia no capitalismo cognitivo. In G. Cocco & S. Albagli (Eds.), *Revolução 2.0 e a crise do capitalism global*. Rio de Janeiro: Garamond.

Albagli, S., & Maciel, M. L. (2010). *Information, power, and politics: Technological and institutional mediations*. Lanham: Lexington Books.

Perroux, F. (1961). *L'économie du XX Siècle*. Paris: Presses Universitaires de France.

Santos, M. (1997 [1996]). *A natureza do espaço: Técnica e tempo. Razão e emoção*. São Paulo: Hucitec.

Santos, M. (2011 [2000]). *Por uma outra Globalização: Do pensamento único à consciência universal*. Rio de Janeiro: Record.

Sartre, J.-P. (1960). *Critique de la raison dialectique*. Paris: Gallimard.

Chapter 4
How Can Santos's Theory and Concepts Help Us to Better Understand Third World Dynamics and Problems?

Aurélien Reys

Abstract During his academic career Milton Santos published approximately 40 scientific works on the so-called Third World. This period of intellectual activity corresponds to the time the geographer was forced to live in an exile that started in France. He mainly wrote about the theme from the point of view of urban agglomerations. To present the reality of these Third World cities, he frequently used figures and empirical studies. He also observed that the dynamism and the expansion of cities in underdeveloped countries depended more on their population growth than on their economy. Milton Santos thought these major cities in the Third World could not be studied as *whole systems* and his observations led to the elaboration of the theory of the *two circuits*. The notion, first introduced in the last chapter of Santos's book *Les Villes du Tiers-Monde*, was further developed in one of his most accomplished works on the theme: *L'Espace Partagé: Les Deux Circuits de l'Economie Urbaine des Pays Sous-Développés*. The author identifies the *upper circuit* as the direct result of a modernization process that drastically changes modes of production and communication, as well as lifestyle and consumption patterns. He understands the *lower circuit* as a consequence of the same modernization processes, but in a more indirect manner: a circuit that includes people and activities that do not benefit at all, or benefit only partially, from recent technological advancements. However, Santos considers these two circuits closely dependent and operating with one another.

4.1 Introduction

Among the several themes Milton Santos approached during his long and illustrious career, some of the most influential and relevant are his works on the so-called Third World. Between the 1960s and the end of the 1970s the Brazilian geographer published approximately 40 scientific articles, books and chapters on the subject,

---

Aurélien Reys, Post-doctoral Researcher, Centre de Coopération International en Recherche Agronomique pour le Développement (CIRAD), France; Email: aurelien.reys@cirad.fr.

© Springer International Publishing AG 2017
L. Melgaço and C. Prouse (eds.), *Milton Santos: A Pioneer in Critical Geography from the Global South*, Pioneers in Arts, Humanities, Science, Engineering, Practice 11, DOI 10.1007/978-3-319-53826-6_4

including two major works: *Les Villes du Tiers Monde* (Santos 1971a, b) and *L'Espace Partagé: Les Deux Circuits de l'Economie Urbaine des Pays Sous-Développés* (Santos 1975).

Interestingly, this period of intense intellectual activity corresponds to the 13 years during which the geographer was forced to live in exile by the Brazilian military dictatorship. Imprisoned because of his proximity to the previous left-wing government and some members of the local Communist party (Lévy 2007), Santos spent 100 days in jail before leaving for France where he had already lived previously for a few years and had completed a doctorate at the University of Strasbourg. France thus came as a natural choice of refuge for the geographer, compounded by the country's academic prestige: France was renowned at the time for its standings in the social sciences and held in particularly high regard by South American academia.[1]

This chapter aims to describe the work of Santos with regards to the Third World. First, it shows how the author's interest in the topic arose and evolved. Then it will present his main contributions by discussing the two main books he published on the subject.

4.2 Milton Santos and the Third World

It is not a coincidence that Santos wrote the chapter that launched his academic career while he was living abroad. Still strongly connected to the African continent, with which they had entertained close economic and cultural relations during a century of colonization, French academics had developed an early and growing interest in issues affecting areas with concentrated poverty.

Nonetheless, Santos did not start to work on the topic during his first stay in France. His doctoral degree—based on a thesis about the downtown area of Salvador —was in fact awarded in the domain of urban geography; it did not have a *Third World*[2] approach. It was only after his return to Brazil and a number of trips abroad, especially to north and sub-Saharan Africa, that Santos began to consider for the first time certain issues related to economic underdevelopment.

---

[1]In the 1930s a group of French academics were in charge of the inauguration and development of the University of São Paulo's (USP) teaching activities. Professors involved included Fernand Braudel (history) Pierre Monbeig (geography), Claude Lévi-Strauss (anthropology) and Roger Bastide (sociology).

[2]The expression *Third World* is used for the first time in 1952 by a French demographer, Alfred Sauvy. Originally, the term was employed in reference to the French 'Third Estate', in order to designate countries neither aligned with the capitalist nor the communist model. At the time when France was still a monarchy, the Third Estate referred to the common people who were part of neither the nobility nor the clergy.

He believed that most of the models used at that time were impossible to apply in impoverished countries. Upon attempting to apply works previously realized by the French geographer Jean Rochefort to the Brazilian state of Bahia, Santos concluded that the hierarchy of urban networks corresponded to factors other than those usually described (Santos 1959a, b). He also came to consider that the overlap between environmental landscapes and human habitats observed in Europe, where evolutions in societies were slow due to a historical stability, was unlikely in the Third World (Santos 2004).

Santos used the term *Third World* for the first time in 1961 in the article 'Quelques problèmes des grandes villes dans les pays sous-développés', published in the *Revue de Géographie de Lyon* (Santos 1961). In the article, Santos couples the term with the concept of *underdevelopment.*[3] Despite his disbelief that developing countries should follow the same trajectory as that of developed countries, he frequently employed the two terms as synonyms throughout his works, like many of his colleagues during the same period. At that time, the expression *Third World* was indeed often overused and regularly interchanged with *underdevelopment* in academic texts. The latter term, underdevelopment, implies a more deterministic vision of progress and the necessary establishment of a market economy.

Although Santos began his work on the topic in Brazil, it was not until his second stay during his exile in France, which lasted seven years, that the geographer explored the Third World theme in more depth. First at the University of Toulouse, then in Bordeaux and Paris, he dedicated most of his research and scientific activity to the improvement of knowledge on the subject, which was still poorly explored and documented at the time.

The debate on the subject of the Third World was, in his opinion, mostly a humanistic discourse about civilization. Being native of the so-called Third World, Santos soon became persuaded that his own vision and understanding of developing countries could contribute to current debates (Santos 2004). He believed that the subject found traction in a postcolonial period in which it was necessary for rich countries to develop academic theories to justify new forms of economic and cultural domination, and this apparent solidarity could accrue profit for the former colonizers (Santos 2004).

4.3 Cities of the Third World

With the exception of a few publications, including the two volumes of *Croissance Démographique et Consommation Alimentaire dans les Pays Sous-Développés* (Santos 1967a, b)—a summary of the courses he gave at the department of liberal

---

[3]'The speed with which cities grow and urban populations increase is a generalized phenomenon in *underdeveloped countries*. This fact is all the more important because it is precisely the *Third World* cities that are the ones that materialize the will of progress and that are preparing the process of development' (Santos 1961: 197).

arts of Toulouse—Santos wrote about the Third World from the point of view of its urban agglomerations. Starting with his first publication on the topic, *A Cidade nos Países Subdesenvolvidos*, he moved beyond comparisons between Europe and Brazil and engaged with case studies of several cities in North and West Africa (Santos 1965a, b). In the following years, many of these studies continued to serve as the basis for Santos's papers and articles, some of which are gathered in the volume *Dix Essais sur des Villes Sous-Développées* (Santos 1970).

In his early studies Santos observed that the dynamism and the expansion of cities in underdeveloped countries depended more on their population growth than on their economy. He also noted that their expansion was sustained both by natural population growth and by migratory influxes issuing from rural areas.[4] The origins of such migration movements were first, according to Santos, a reflection of a social change:

> Those people invading the streets and overpopulating slums and shantytowns are rather attracted by the city than expelled and chased away by a rural area unable to feed them. The city has no jobs to offer. But there is so much difference between the city and its hinterland that the peasant looks at it as an opportunity to get a better life and live in a place with higher standards. Being a pariah in the countryside or a pariah in the city, between the two options the peasant prefers the second one. As described by the Brazilian novelist Jorge Amado in a book dedicated to the city of Salvador, he comes to "take part in the show". (Santos 1961: 201)

Santos frequently used figures and empirical studies to present the reality of these Third World cities and, after a decade of works on the theme, gathered the results of his analysis in a new book entitled *Les Villes du Tiers Monde*[5] (Santos 1971a, b). This work aimed to present in depth the different characteristics of cities found in the developed and non-developed worlds. Santos drew attention to the fact that Third World cities shared several traits in common. Apart from their demographic growth, they were also characterized by features such as: a significant disproportionate relationship between their total and economically active inhabitants; profound social inequality of living standards; and a generally more youthful population. He also pointed out numerous other elements that differentiated the cities analyzed, such as their geographical locations, size and original function in national economic production.

Ultimately, at a time in which the debate between liberals and Marxists was raging, Santos did not miss the chance to position himself in favour of the latter, offering an analysis consistent with unequal exchange and dependency theories.[6] Santos regularly accused colonial industries of crippling urban growth in Third World cities (Santos 1971a, b), and emphasized the city's role in exchanges that

---

[4]He also pointed out in an article on downtown Salvador that the city was not growing due to dynamism, but because of its lack of dynamism (Santos 1958).

[5]Book first published in French.

[6]Dependency theory is a thesis created by Samir Amin and Arghiri Emmanuel, which argued that underdevelopment is a consequence of historical processes resulting from an economic dependency that deteriorated the terms of trade to the disadvantage of poorer countries.

disadvantaged suburban areas, an analysis similar to that of *centre-periphery* relations between the First and the Third World:

> What characterizes the largest cities of underdeveloped countries is their role as bridges between the industrial world that buys their commodities, and the rural world that provides raw materials, the latter of which, in return, receives manufactured goods produced or imported by those cities. (Santos 1961: 198)

Santos finally pointed out the disconnection between these cities, organized in favour of foreign interests, and their hinterlands. According to him, relations between cities and their hinterlands created an imbalance in social living conditions for the populations of these two areas. Such inequalities lead to the creation of slums or favelas illegally built on cheap land where construction is often extremely difficult and dangerous. Santos noted that most of these slums, often located at the city's margins but also found in city centres, were progressively becoming permanent housing for under-skilled workers and newcomers. He also stressed that, contrary to the most developed countries, the urbanization in the Third World was mainly structured by population growth and flows of migrants (Santos 1971a, b).

4.4 The Two Circuits and the Shared Space

Milton Santos thought these major cities in the Third World could not be studied as '*whole systems*' (Santos 1974: 276) and in the late 1960s, building upon economist Arthur Lewis's perspective on dualism in the labour market, he became interested in informal urban economic activities. These included diverse low-paying jobs such as car guards and street vendors. Using a concept coined by French geographer and friend Jacqueline Beaujeu-Garnier, Santos codified such activities as comprising a *primitive tertiary sector* (Santos 1968), whose elasticity was an important driving force behind the developing economy.

These observations led to the elaboration of Santos's theory of the *two circuits*, probably his most original contribution to the study of Third World. The notion, first introduced in the last chapter of Santos's book *Les Villes du Tiers-Monde* (Santos 1971a, b), was further developed in one of his most accomplished works on the theme: *L'Espace Partagé: Les Deux Circuits de l'Economie Urbaine des Pays Sous-Développés* (Santos 1975).[7] The latter represented an important turn in Santos's academic research, as the author started to devote more attention to developing theoretical concepts rather than analyzing empirical data.

Santos noted that in underdeveloped countries space is constantly reorganized through sociologic and economic forces that are responsible for creating significant differences in income. Those forces tend to impose, within the same area, coexisting activities of a same sector but taking place at different stages. These activities are

---

[7]Until the recent publication of Toward an Other Globalization, this was the only book of Santos's translated into English (see Santos 1979).

related to the variety of individual patterns of consumption and therefore include both modern and traditional services. Santos argued that such a situation leads to the creation of two different economic circuits that, nevertheless, represent different sides of the same phenomenon stemming from modern technology (Santos 1975).

What Santos identified as the *upper circuit* is the direct result of a modernization process that drastically changes modes of production and communication, as well as lifestyle and consumption patterns. This circuit consists of activities created by technological progress and the population that benefited from such advancements. Santos also pointed out that the technologies used are primarily modern, imported and have an influence on the organization of space at a large scale (Santos 1975). This circuit employs a significant number of foreigners and sets fixed prices for goods and services. The goals of production are long term and aim to perpetuate growth through the accumulation of capital. Santos considered that this circuit consists of 'impure' industry and export activity, serving only outsiders' interests and installed in underdeveloped cities only because of the comparative advantages offered by the latter's respective locations (Santos 1975).

Santos also understood the *lower circuit* as a consequence of the same modernization processes, but in a more indirect manner. This circuit includes people and activities that do not benefit at all, or benefit only partially, from recent technological advancements. Manual labour predominates and is conducted in a restricted amount of space. In contrast to the ones belonging to the upper circuit, the activities forming the lower circuit are inadequate to transform the organization of space at a large scale (Santos 1975). The circuit is also regarded as employing local or poor immigrants and including traditional production such as detailed handmade objects. Bargaining is common and fluctuations in prices are substantial. In opposition to the upper circuit, the lower circuit responds to short-term objectives and accumulates little capital. Santos stated that this lower circuit nonetheless possesses a social and economic role, as it constitutes a welcoming structure for people migrating from rural to urban areas.

The two circuits are not isolated from one another. On the contrary, they are closely dependent and operate with each other. Additionally, activities belonging to one circuit can sometimes partially present some of the other one's characteristics. This phenomenon is especially noticeable within the upper circuit to the extent that it inspired Santos to codify the concept *marginal upper circuit*. Such activities that present features of the lower circuit are particularly common in the case of regional cities that maintain, compared to other large metropolitan areas, tighter-knit interactions with the inferior circuit and the population it involves. In effect, then, Milton Santos observed that the relations between the two circuits vary according to the importance of the city: while the influence of the lower circuit is confined to the limits of major urban areas, it extends beyond urban boundaries in smaller towns. The various sizes of the demand market can be seen as a primary reason for such differences.

Toward the end of his career the Brazilian geographer came to the conclusion that the way geographic space is organized in Third World cities, between rich and poor areas, is a supplementary reason for the polarization of the economy (Santos 1975). Spatial organization reinforces the duality of the economy and its separation into two different circuits, as efficiency of transport varies from one place to another and does not play a fully integrating role. Furthermore, new means of communication tend to put the most vulnerable at the mercy of a consumption model that benefits the more privileged groups.

Santos did not, however, criticize the urbanization process. On the contrary, he considered an increase in the number of major cities and the implementation of industries oriented toward endogenous needs to be desirable. He assigned the state an important role in the urbanization strategy to eliminate social disparities and a dual functioning of the economy. Nevertheless, he also stated that the lower circuit would certainly continue to exert an important influence during this period of transition, due to the relative slowness of social transformation processes. Santos insisted, though, that this situation should not become permanent (Santos 1975).

Forty years later, despite the decrease of the population growth rate and the flux of internal migration, it seems that, as Santos predicted, the lower circuit is far from disappearing. As in numerous other developing countries, the informal sector has continued to grow in Brazil and still bears a remarkable resemblance to that described by the geographer in the 1960s and 1970s, thus making Santos's work pertinent for analyzing contemporary urbanization.

4.5   Conclusion

Considering the geopolitical context of the time, as well as the geographer's origins, it appears logical that Milton Santos dedicated the first part of his career as a researcher to theorizing the two circuits in the Third World. Coming from a relatively modest area in the Brazilian state of Bahia, Santos had a vision of the world that was very different from most of his French colleagues. As he noted in the introduction to his last book on the theme written in French, *L'Espace Partagé*:

> I was privileged in this long and patient research, as I belong to the Third World, having been able to travel to many Latin American and African countries, and having talked to many people from the Third World, whether they be theorists or simple fellows with the daunting task of facing reality. (Santos 1975: 3)

However, if Santos's approach can be considered particularly relevant because it fits within a framework of lived experience, this same frame appears to have also restricted his view. For instance, most of the time his work uses examples of European cities in comparison with those of the Third World, and rarely considers North American examples. He also did not focus on the reality of Asian cities, which have played a major role in the economic and social transformations of the world; their influence within international markets has indeed come to challenge former hierarchies that Milton Santos often referred to as being unsurpassable.

Santos's work on the Third World remains, however, a major advancement for geography and the understanding of urban issues in countries facing important socio-economic challenges. Undeniably, Milton Santos stands as a major contributor to numerous constantly evolving issues, which remain far from being resolved.

References

Lévy, J. (2007). *Milton Santos: Philosophe du mondial, citoyen du local*. Lausanne: Presses Polytechniques et Universitaires Romandes.
Maurel, J. (1996). Homenaje al profesor Milton Santos. *Anales de Geografía de la Universidad Complutense, 16*, 203–223.
Santos, M. (1959a). *Le centre de la ville de salvador: Etude de geographie urbaine*. [Ph.D. dissertation]. University of Strasbourg I, Faculty of Geography and Management.
Santos, M. (1959b). Quelques problemes geographiques du centre de la ville de salvador. *L'Information Géographique, 23*, 93–98.
Santos, M. (1960). Geografia e desenvolvimento econômico. In M. Neto & A. Luís (Eds.), *Desenvolvimento: Problemas e soluções* (pp. 107–126). Salvador: Imprensa Oficial de Bahia.
Santos, M. (1961). Quelques problemes des grandes villes dans les pays sous-developpes. *Revue de Géographie de Lyon, 80*, 197–218.
Santos, M. (1965a). Villes et region dans un pays sous-developpe: L'exemple du roncocavo de Bahia. *Anales de Géographie, 406*, 678–694.
Santos, M. (1965b). *A cidade nos países subdesenvolvidos*. Rio de Janeiro: Civilização Brasileira.
Santos, M. (1966). Vues actuelles sur le probleme des bidonvilles. *L'Information Géographique, 30*(4), 35–42.
Santos, M. (1967a). *Croissance demographique et consommation alimentaire dans tous les pays sous-developpes, 1. Les données de base*. Paris: CDU.
Santos, M. (1967b). *Croissance demographique et consommation alimentaire sans tous les pays sous-developpes, 2. Milieux géographiques*. Paris: CDU.
Santos, M. (1968). Le role moteur du tertiaire primitif dans les villes du tiers monde. *Revista do Instituto Geografico e Histórico da Bahia, 37*(2), 1–16.
Santos, M. (1969a). *Aspects de la geographie et de l'economie urbaine des pays sous-developpes*. Paris: CDU.
Santos, M. (1969b). Alimentation urbain et planification regionale en pays sous-developpe. *Tiers-Monde, 37*, 95–114.
Santos, M. (1970). *Dix essais sur les villes des pays sous-developpes*. Paris: Éditions Orphys.
Santos, M. (1971a). *Les villes du tiers monde*. Paris: M.-Th. Génin, Librairies Techniques.
Santos, M. (1971b). *Le metier de geographe en pays sous-developpe: Un essai methodologique*. Paris: Éditions Orphys.
Santos, M., & Kayser, B. (1971). Espace et villes du tiers monde. *Tiers-Monde, 12*(45), 7–13.
Santos, M. (1972). Dos circuitos de la economia urbana de los paises subdesarrolados. In J. C. Funes (Ed.), *La ciudad y la region para el desarrollo* (pp. 67–99). Caracas: Comision de Administracion Publica de Venezuela.
Santos, M. (1973). *Underdevelopment and poverty: A geographer's view*. Toronto: University of Toronto.
Santos, M. (1974). Sous-développement et pole de croissance economique et sociale. *Tiers-Monde, 58*, 271–286.
Santos, M. (1975). *L'espace partagé: Les deux circuits de l'economie urbaine des pays sous-developpes*. Paris: M.-Th. Génin, Librairies Techniques.

Santos, M. (1977). Spatial dialectics: The two circuits of urban economy in underdeveloped countries. *Antipode, 9*(3), 49–60.

Santos, M. (1979). *The shared space: The two circuits of the urban economy in underdeveloped countries*. New-York: Methuen.

Santos, M. (2004). *Testamento intelectual*. São Paulo: UNESP.

Chapter 5
Psychosphere and Technosphere: Complex Relations in the Hospital Realm

Eliza Pinto de Almeida

Abstract The improvement in the health profile of populations is strongly associated with investments in technological innovations. Hospitals are fixed elements that constantly incorporate these innovations. The objective of this chapter is to analyze the metamorphoses that has taken place in different time periods leading to changes in the psychosphere and technosphere of hospitals. As demonstrated by Michel Foucault, until the 18th century hospitals were not seen as a place of cure and a psychosphere of fear prevailed. It was only around the 1780s that hospitals became a therapeutic instrument. In the technical-scientific period, thanks to new scientific discoveries and new inventions that were incorporated into diagnosis and treatment, the hospital technosphere formed and strengthened a favorable psychosphere. In the current technical-scientific-informational period, there is a complex hospital technosphere that incorporates innovations from the medical/hospital equipment industry, as well as new drugs and research centres. The hospital psychosphere reflects approval of this constant technical-scientific incorporation and, as a consequence, hospitals have become a fixed element that is valued by society as a whole.

5.1 Introduction

Hospitals are highly valued technical and scientific objects in contemporary societies. In the social imaginary they are often associated with the on-going incorporation of new technologies, state-of-the-art therapies and highly skilled professionals. They are thought of as a privileged *locus* of cure. As a result, these institutions help establish a psychosphere in which hospital care manifests as a core value. According to Milton Santos, the psychosphere is 'the result of beliefs, whishes, and habits that inspire philosophical and practical behaviors, interpersonal relationships, and the communion with the universe' (Santos 2008: 30). Intrinsically linked to the psychosphere is the technosphere, a realm that stems from a growing

Eliza Pinto de Almeida, Associate Professor, Federal University of Alagoas, Brazil; Email: eliza. almeida@igdema.ufal.br.

© Springer International Publishing AG 2017   55
L. Melgaço and C. Prouse (eds.), *Milton Santos: A Pioneer in Critical Geography from the Global South*, Pioneers in Arts, Humanities, Science, Engineering, Practice 11, DOI 10.1007/978-3-319-53826-6_5

artificialization of geographical space and adapts to the commandments of production and exchange (Santos 2004). At of the end of the 19th century the psychosphere and technosphere created in and through hospitals underwent deep changes, fruit of the transformation of geographical space itself. This chapter aims to briefly reconstruct some of the interactions that led to the transformation of hospitals and their interconnection with spatial dynamics. In order to do that, it shall establish a periodization that reveals the metamorphoses undergone by the psychosphere and technosphere of hospitals. From the Middle Ages to the 18th century, hospitals were not meant to cure, as Michel Foucault states in *The Birth of the Clinic*, and a psychosphere of fear prevailed. It was only in the late 18th century, more specifically in the 1780s, that hospitals appeared as therapeutic instruments. In the technical-scientific period, due to new scientific discoveries and new inventions that were incorporated into diagnosis and treatment, the hospital technosphere formed and strengthened a favorable psychosphere. In the current technical-scientific-informational period, there is a complex hospital technosphere that incorporates innovations from the medical/hospital equipment industry, as well as new drugs and research centres, among others. The hospital psychosphere reflects satisfaction with this constant technical-scientific incorporation and, as a consequence, hospitals have become a fixed element that is valued by society as a whole.

Santos (2008) argues that space is shaped by both fixed and flowing objects. Fixed objects refer to physical entities such as post offices, bank branches, hospitals and factories. Flowing dimensions are movements conditioned by actions. In the case analyzed herein, hospitals are fixed elements usually located in urban areas. They are, therefore, part of the logic of the organization of cities. Flowing elements directly connected to hospital activities include patients, medical staff, nurses, technicians, as well as management, cleaning and maintenance personnel, among others. Fixed and flowing dimensions of space interact and transform each other, creating an 'indissoluble, solidary and contradictory set of systems of objects and systems of actions that cannot be considered in insolation, but as the unique frame in which history unfolds' (Santos 1996: 51).

On the one hand, the hospital technosphere articulates different fixed and flowing elements that directly or indirectly support hospital functioning. These range from clinical labs, diagnostic centres, doctors' offices and healthcare research centres to specialized trade, since hospitals are important centres of consumption that generate heterogeneous demands from the medical/hospital equipment industry, from the pharmaceutical industry and from a myriad of small and medium manufacturers that regularly supply hospitals with products used in routine procedures (like gloves, reagents, disinfectants, etc.). In many countries, health management organizations (HMOs) may also be added to the list due to their increasingly close relationships with hospitals. On the other hand, the hospital psychosphere is driven by the trust that the majority of the population places in this fixed element as an important centre for rehabilitation and cure (and that incorporates the latest technologies available in the market).

Santos's frame of fixed and flowing objects, and of psychosphere and technosphere, is thus useful for analyzing the medicalization of hospitals: in the 18th century, technical and scientific advances were absorbed by, and transformed, these institutions. Since then, there has been a confluence of diverse interests involved in hospital care that is conditioned by the transformation of geographical space via both its fixed and flowing elements.

However, as described below, the hospital technosphere has not always been one of trust. In the Middle Ages and throughout most of the Modern Age, from the 15th to the 18th centuries, hospitals were a place of isolation where the poor and in-need would wait for death. The first institutional metamorphosis took place after the Industrial Revolution, in the second half of the 18th century, when hospitals became medicalized and provided with a specialized medical staff, thus creating a more complex fixed dimension. Although they became a place of cure and care for the ill, a paradigm shift about hospitals had not yet taken place in the social imaginary of the population and they continued to be seen as a fixed element focused on the poor and in-need. It was only after the advent of *technological* hospitals—in which scientific and technical advances were increasingly incorporated into medical practice and into the hospital environment itself, a byproduct of the scientific and technical revolutions—that hospitals achieved the primacy observed in contemporary societies. Fed by a technosphere that incorporated state-of-the-art scientific and technological resources, this new psychosphere valorized hospitals, which began to provide care for people from all social strata. In the current technical-scientific-informational era[1] (Santos 1996), hospitals have become the nerve centre of healthcare systems.

5.2 The Psychosphere and the Technosphere of Fear: Hospitals in the Middle Ages

The concept and form of hospitals have changed dramatically throughout history. In ancient Egyptian and Greek civilizations, fixed dimensions of geographical space, such as sanctuaries and temples, served to care for the ill and the wounded. Examples include the temple of Imhotep, in Egypt, and the temples dedicated to Asclepius, in Greece. It was within these fixed elements, societies believed, that sleep coupled with remedies prepared by priests could reestablish the health of the ill (Campos 1944).

The rise of Christianity led to the first hospitals maintained by religious orders, which provided shelter to travelers as well as to ill, poor and elderly people. Hospitals, however, were not seen as a place to cure the ill. They served the purpose of providing the ill and the poor with assistance while they also ensured that these

---

[1]According to Santos (2008: 123), the technical-scientific-informational period is 'marked by new signs, such as: the multinationalization of firms and internationalization of production and products; generalization of credit, which reinforces the characteristics of economization of life; new state roles in a globalized society and economy; a circulation frenzy that has become an essential factor of accumulation; the great information revolution that instantly connects places, thanks to advances in computing'.

people were excluded and separated from the rest of the society. Foucault noted that medieval hospitals were not medical institutions, as patients were taken care of by lay and religious people whose mission was to 'save the soul of the poor in the moment of death' (Foucault 2003: 102).

The Church, as the most powerful and universal institution in medieval civilization, built an increasing number of hospitals to provide general care for the sick and weak. In the Middle Ages the Church, feudal lords and the nobility sustained the feudal system. In Europe, from the 12th century on, hospitals began to spread as a function of the growth of cities. Human agglomeration and the rise of trade routes propagated different diseases such as measles, smallpox, tuberculosis and leprosy. The number of hospitals grew so as to isolate the ill. Some hospitals were even specialized: leprosariums, for example, isolated leprosy patients, while lazarets were quarantine stations for people coming from regions struck by the bubonic plague in the 14th century.

In 16th century Europe it was increasingly clear that the feudal world was breaking apart and making way for capitalism and the bourgeoisie. Hospitals, at this moment, broke away from religious orders, but kept their social assistance character. This began to change at the beginning of the 18th century when hospitals became central to the regulation of urban life, segregating and isolating all of those who presented a threat to society. Hospitals were places where the sick, the mad, perverts and prostitutes, amongst others, were confined (Foucault 2003).

From the Middle Ages to the 18th century the psychosphere of hospitals was one of apprehension and fear due to the high mortality rates resulting from the lack of knowledge about disease transmission and lack of hygiene in hospital environments. As previously stated, hospitals were not seen as an environment of cure. In Michel Foucault's words, 'the hospital was a place where one went to die' (Foucault 2003: 102). This psychosphere, in which hospitals are understood as a place to die, began to shift in the second half of the 18th century. For Foucault (2003), a clear understanding that hospitals could and should be instruments of cure arose around 1780; this new approach was signaled by a new practice based on systematic observations by physicians. According to Rego (1993), the emergence of *therapeutic* hospitals was a result of the institutions' medicalization. In other words, hospitals were reorganized due to the valorization of the therapeutic act, which became a major purpose of their activities and began to shape a new psychosphere.

Medicalization reflected larger social transformations. On the heels of the Scientific Revolution in the 15th century, and especially after the Industrial Revolution in the 18th century, a new mindset—increasingly distant from religious precepts—was forged. Such new social arrangement played an important role in the institution of hospitals, which, after being medicalized, became a space of study, investigation, treatment and medical training.

This was new: prior to the Enlightenment, physicians did not have systematic knowledge of the cases they treated. The medical practice was essentially based on intuition. Advances in science, which began in the Renaissance, started to impact other fields of human knowledge such as medicine, and these fields were gradually transformed. One of these transformations was due to the advent of surgery in the

18th century (Abreu 2007). This had significant impacts on medicine and, consequently, hospitals. Starting in the 18th century, medical training became a part of the routines found in hospitals and the hospital acquired the status of a place where therapeutic actions were taken against illness and disease.

The early 18th century saw another paradigm shift: the deterioration of an individual's health became attributed to the internal structure and (dis)organization of hospitals. Foucault (2003) tells us that research commissioned by the Academy of Science and conducted by French physician Jacques Tenon and Englishman John Howard in hospitals, prisons and lazarets in Europe between 1775 and 1780 revealed a connection between a hospital's internal organization and the mortality rate of its patients and thus incited transformations in hospital architecture. There was, as a consequence, a growing concern with controlling flows in the hospital environment. More medical attention was devoted to the physical surroundings of ill people: air, water, temperature and diet had to be controlled. Medical knowledge of the time believed in miasmas—that diseases could be transmitted by odors. As such, putrid water and waste had to be eliminated. Architecture, then, became a crucial element in creating new fixed dimensions of hospitals, ones that were deemed vital for the curing process:

> Hospital architecture becomes an instrument of cure in the same category as a dietary regime, the practice of bleeding or other medical actions. The space of the hospital is medicalized in its purpose and its effects. This is the first characteristic of the transformation of the hospital at the end of the eighteenth century. (Foucault 2003: 109)

Concerns of the time also included the proximity of nurses treating wounded people to those of women giving birth, and the cross contamination of materials containing pathogens such as clothing, sheets and rags used to treat wounds. The internal structure of hospitals changed according to the specialized activities they housed. As a consequence, hospital flows were controlled insofar as people with contagious diseases, women in labour and other patients were placed far from each other. The hospital's fixed dimension therefore changed as a function of this new architecture, which aimed to end unhygienic environments that lacked ventilation and sunlight that were propitious for contamination.

Concurrently there was a growing attention to detailing the clinical conditions of each patient and the medical procedures used. Foucault (2003: 111) states: 'I understand by 'clinic' [la clinique] the organization of the hospital as a place of formation and transmission of knowledge [savoir].' This was the beginning of the *rationalization* of hospital administration, with the making of permanent, detailed and individualized records for each patient. Physicians became major figures in this new hospital configuration. They now searched for explanations of what caused disease, rather than simply intervening in the crises that disease provoked. The hospital had shifted from being a place of charity, spiritual salvation and/or social segregation to being a curing machine.

The combination of technical advances in constructing hospitals and the presence of physicians in these institutional structures shaped a new technosphere. This shift also entailed a gradual change in the psychosphere. There was a decline in the negative connotations of former hospital spaces, which were fearfully thought of as sites of disease transmission and as places that could threaten the surrounding population. In the social imaginary, hospitals became associated with the cure of and care for the poor.

5.3 The Hospital's Technosphere and Psychosphere in the Scientific and Technical Period

In the last quarter of the 19th century the world witnessed the breathtaking expansion of the capitalist system. The Second Industrial Revolution ushered in a new scientific and technical era which was 'felt to its fullest, changing habits and daily customs, as well as the pace and intensity of the means of transportation, communication and work' (Sevcenko 2009: 11–12). The modernization of engineering systems dramatically changed the geographical environment, allowing for new flows of people, merchandise and ideas. Scientific advances were also manifest in the chemical industry, as well as in the new branches of the metal and the steel industries and allowed for the exploration of new forms of energy such as electricity and oil derivatives. Here, 'the technical objects and the technological space are superior *loci* of actions, as they successfully override the natural forces' (Santos 1996: 189). The set of changes in the geographical environment would have far-reaching impacts on the medical and hospital system.

In this period the medical/hospital technosphere incorporated, in an unplanned manner, advances resulting from new scientific discoveries especially in the biomedical field. For instance, specific therapies were developed (such as radiation therapy for cancer), as well as inventions that were slowly incorporated into patients' diagnosis and treatment. Physicians created a consumption demand for therapeutic technologies, leading to a permanent production of hospital materials. Additionally, state and philanthropic institutions, medical societies and research centres required further knowledge about the treatment of specific diseases, urging states to invest more capital and hospitals to increasingly specialize. However, there was no systemic articulation between different fixed and flowing dimensions that became part of this hospital technosphere because of the limitations of the technique[2] itself. Today, to the contrary, geographical objects that comprise the hospital technosphere articulate in systems amongst themselves, as we will see below.

Many technical and scientific advances were incorporated into medical practices in Europe. In 1842 Crawford Long (1815–1878) discovered the anesthetic effect of ether, which led to its widespread use in surgical procedures. French doctor René

---

[2]For Milton Santos, techniques are a set of instrumental and social means by which people live their lives, produce and, at the same time, create space: 'In whatever fraction of space, each variable reveals a technique or a set of particular techniques. One may also say that the functioning of each one of those variables relies precisely on those techniques. Taking world history as a reference, every technique may be located in time' (Santos 2008: 61).

Laennec invented the stethoscope, a device that enabled doctors to listen to the heart beat of patients. French chemist Louis Pasteur (1822–1895) spearheaded studies on contagious diseases, which demonstrated that putrefaction and fermentation processes were associated with the presence of microorganisms. His work revolutionized medical concepts, laying the foundation for modern microbiology and a germ theory of disease.

Influenced by Pasteur, English surgeon Joseph Lister (1827–1912) applied germ theory to destroy microorganisms from wounds and surgical incisions using phenic acid as an antiseptic and significantly decreasing mortality associated with postoperative infection (Tubino & Alves 2009). Sterilization—leading to asepsis—was gradually accepted by physicians and adopted as an important preventive measure. Charles Chamberland, an associate of Pasteur, developed the autoclave, the first steam sterilizer; achieving temperatures over 120 °C, it represented an important step toward the disinfection of devices used in medical practice. By the beginning of the 20th century the systematic use of the autoclave to sterilize medical instruments and combat microbial infections in hospitals was widespread. The use of surgical gloves in the operating room was another important measure for asepsis. They were created in 1889 at Johns Hopkins Hospital, Baltimore, USA, and had a positive impact on infection rates (Tubino & Alves 2009). Antisepsis and asepsis techniques, in general, enhanced the control over the hospital environment.

Diagnostics were also improved with the incorporation of new devices into medical practice. Once again, Louis Pasteur played a central role by stimulating the use of microscopes to perform laboratory diagnostic tests and investigate the action of microorganisms. The discovery of the X-ray (1895) by German physicist Wilhelm Conrad Rontgen also attracted great interest, especially among physicians, since, for the first time, the inside of human body could be seen.

A number of different technological advancements marked the hospital as a site of treatment. In the late 1910s, in the USA and Europe, the number of cancer patients was increasing due to the longer life span of the general population. As a result, new hospitals specializing in cancer treatment were created that applied radiation in therapeutic form (Teixeira 2010). In 1928 the discovery of penicillin by English doctor Alexander Fleming further revolutionized medical treatment, making it possible to control infectious disease. By 1940 penicillin was widely used as an antibiotic. The adoption of these new therapies was enabled by technological advancements in the manufacture of X-rays and large-scale industrial production of penicillin, epitomizing the intrinsic relation between new technologies and medical practices.

These advances were central to the creation of the technological hospital of the late 19th century, which impacted its psychosphere. The hospital technosphere had become significantly more complex than it had been in previous eras. It was now associated with scientific rationalism and with new diagnostic and therapeutic technologies, both strongly related to industrial development (Braga 2000). The general public, in turn, began to identify the hospital as an important place of cure and treatment: 'For the first time in history, hospitals started being used... by the entire population, becoming the setting where sanitary happenings in the life of a human being took place, from birth to death' (Neufeld 2013: 13).

Based on the technical and scientific advances associated with industrial development, then, the new technosphere consolidated the psychosphere of the hospital; the institution became increasingly valued in society as it became understood as a *locus* of cure.

As a result of these shifts, the technological hospital had acquired a large number of fixed and flowing elements. These went beyond the hospital's walls, reaching laboratories, the burgeoning pharmaceutical industry and the chemical industry. Hospital-related fixed and flowing elements did not yet have a systemic functioning, but the first half of the 20th century consolidated the interdependence of several types of industries and services. In 1904 Siemens-Reiniger industries produced the first X-ray machine (Navarro 2009). Throughout the 20th century radiation therapy was developed due to the articulation of several areas of medicine, technology, physics and biology. The pharmaceutical industry, in turn, developed a series of new drugs that were incorporated into therapeutic practice.

This synergy transformed the psychosphere and technosphere of medicine generally and hospitals specifically. The dissemination of new knowledges and technologies has continued until the present moment. It is worth noting, however, that the specific geographical *place* is what endows medical technologies with 'the principle of historical reality, relativizing their use, integrating them into a set of life, removing them from their empirical abstractions and attributing historical effectiveness to them' (Santos 1996: 48). In other words, the incorporation of medical technologies into medicine has been conditioned by spatial arrangements.

5.4 The Complex Hospital Technosphere and Psychosphere in the Technical-Scientific-Informational Period

The technical-scientific-informational period, to Santos, represents the peak of the internationalization of capitalism resulting from economic and financial globalization. Information is the key variable at present and its rise is directly connected to progresses in computing, robotics, telecommunications, chemistry, biotechnology, genetic engineering and many other fields that have participated in productive systems to different extents from the 1970s on. Technique, science and information are selectively diffused in geographical space by means of new systems of informational objects and actions that meet the needs of a globalized market.

In this period—from the second half of the 20th century until the present moment—the provision of healthcare services has become increasingly dependent on a complex medical/hospital framework. As Schraiber (1993) describes, the development of science and technology gained momentum, which resulted in the increasing specialization of medical disciplines; physicians had to learn the ever-evolving diagnostic and therapeutic advances in their respective specialties. Training institutions also used a complex set of instruments that required specialized knowledge (Pereira 2003).

It is possible to argue that hospitals are, today, the nerve centre of the healthcare-related technosphere. They are central to the medical/industrial complex, which articulates medical care, professional training networks, the pharmaceutical industry, and the medical device and diagnostic instrument manufacturing industries. There are multiple interests—including many academic and market interests—involved in the dynamics of the medical/industrial complex. Large transnational companies, for example, deploy myriad strategies that boost innovation and competitiveness in the medical sector. According to Gadelha and Augusto (2003), the health sector is one of the sectors with the highest degree of interaction between universities, research institutes and the business community. The hospital technosphere during the technical-scientific-informational period encompasses diverse objects and actions related to medical/hospital care that articulate amongst themselves into coherent systems in a way not realized in the former periods.

The pharmaceutical industry has emerged as a key player in consolidating the technosphere during this current period. In the 1950s it began to provide new drugs capable of solving 'old' health problems such as medication for diabetes and hypertension, antidepressants and others. Physicians and patients believed the drugs produced by this industry to be one of the 'great accomplishments of humanity' (Vianna 2002: 379). Landim et al. (2013) note that the global market in medical equipment is highly concentrated. Although these companies have a worldwide presence, 50% are North American, 30% European, and 10% Japanese. The authors also state that, in the last decade, the global medical equipment market doubled in size, netting approximately US$ 325 billion in 2011. In parallel, the capitalist sector is also represented by the large equipment and diagnostic industries, which increase doctors' reliance on new technological apparatuses. Hospital services are thus among the most dynamic sectors that drive and absorb new market-driven technologies.

As part of their strategy, pharmaceutical companies became increasingly present in the academic sphere. Luz (1979) demonstrates that, in the 1950s in Brazil, the influence of pharmaceutical labs in hospital care manifested in the academic professor who develops and publishes investigations sponsored by these great laboratories (usually studies proving the efficacy of certain drugs): 'this is the professor who innovates in teaching practice, endowing it with the features of the new reality of medical knowledge: specialized and intertwined with the pharmaceutical industry' (Luz 1979: 176).

Chemical and biotechnological industries, which create vaccines, diagnostic reagents and hemoderivatives, are also central to the medical/industrial complex. Basic physics and mechanical, electronic and material engineering advancements have created new equipment and instruments, such as orthoses and prostheses, and general consumption materials that play a significant role in innovation. The importance of the microelectronics revolution has been crucial here because it has sped up technological innovations in a multiplicative fashion.

Diagnostic imaging, in turn, has become more sophisticated with the appearance of tomography and magnetic resonance scans. These scans have many applications in different medical specialties such as cardiology, oncology, neurology and dermatology. The production of these scans is centralized: three major manufacturers— GE Healthcare, Philips, and Siemens—control 75% of the world market (Landim et al. 2013). Such centralization of production increases the costs of medical/hospital care, with significant impacts on public healthcare systems because a large portion of society cannot afford it. At present, this centralization of production strengthens the hegemonic action of large enterprises and intensifies competitiveness among them. For Santos (2008: 35): 'in the current days, competitiveness as a discourse occupies the place that was once occupied by Progress—in the beginning of the century—and by Development—in the post-war period. Prior to this, however, the debate was a philosophical, teleological one'. Today, the competitiveness-based debate does not need an ethical justification, 'just like any other form of violence, for that matter' (Santos 2008: 35). In healthcare, the centralization of production of medical/hospital equipment and drugs, among other products, is dominated by a small number of companies that fiercely compete for the global market.

The improvement in the health profile of the population is increasingly associated with investments in technological innovations. This has created a powerful psychosphere that associates hospitals with cure, treatment and care. This psychosphere has been constituted through the health-related technosphere and is underpinned by the medical/industrial complex, which is shaped by different economic and financial interests. Hospitals are consumers of industrial products, equipment and instruments, and pharmaceutical products. They are central to the development of and profit in those industries. Hospitals are also part of a technical apparatus comprised of laboratories, healthcare insurance providers and maintenance service providers. This constant growth of the technological frontier results in an increase in healthcare expenditure (Landim et al. 2013).

The quality of healthcare depends on next generation equipment, diagnostic tests and prescription drugs. The demand for these materials comes from physicians and patients alike because the hospital's psychosphere is now associated with cure and care. Technological advancement here is understood as human knowledge being applied to its full potential to solve individuals' health problems (Vianna 2002).

Nevertheless, the constant incorporation of technology creates a permanent tension in healthcare systems. The strict relations established by the hospital complex tend to reduce healthcare to a mere economic sector that obeys market rules and is relentlessly attempting to obtain growing profits. The risk of market-oriented healthcare systems increases as the process of economic globalization advances: the logic in which nations must operate is one that serves the interests of the markets, thus abandoning social classes that cannot afford the high costs. Santos (2000: 58) pointed to the social costs of this phenomenon: 'the abandonment of the idea of solidarity lies behind this understanding of economy and leads to the situation of helplessness in which we live today'. In today's world, solidarity is replaced by perversity. The logic of profitability that feeds the medical/hospital technosphere and psychosphere demonstrates this perversity, which is gaining momentum in the contemporary era.

References

Abreu, J., & Luiz, N. (2007). Os estudos anatômicos e cirúrgicos na medicina portuguesa do século XVIII. *Revista da SBHC, 5*(2), 149–172.

Braga, D. (2000). *Acidente de trabalho com material biológico em trabalhadores da equipe de enfermagem do centro de pesquisas hospital evandro chagas.* National School of Public Health: Oswaldo Cruz Foundation.

Campos, E. S. (1944). Breve notícia histórica sobre os hospitais em geral. *História e evolução dos hospitais* (pp. 7–46). Rio de Janeiro: Ministério da Saúde.

Foucault, M. (2003 [1979]). *Microfísica do poder.* Rio de Janeiro: Graal.

Foucault, M. (2015). The incorporation of the hospital into modern technology. In J. W. Crampton & S. Elden (Eds.), *Space, knowledge and power: Foucault and geography.* London: Ashgate.

Gadelha, C., & Augusto, G. (2003). O complexo industrial da saúde e a necessidade de um enfoque dinâmico na economia da saúde. *Ciência & Saúde Coletiva, 8*(2), 321–535.

Landim, A., Gomes, R., Pimentel, V., Reis, C., & Pieroni, J. P. (2013). Equipamentos e tecnologias para saúde: Oportunidades para uma inserção competitiva da indústria brasileira. *BNDES Setorial, 3,* 173–226.

Luz, M. T. (1979). *As instituições médicas no brasil: Instituições e estratégia de hegemonia.* Rio de Janeiro: Graal.

Navarro, M. V. T. (2009). Evolução tecnológica do radiodiagnóstico. *Risco, radiodiagnóstico e vigilância sanitaria* (pp. 31–36). Salvador: EDUFBA.

Neufeld, P. M. (2013). Uma breve história dos hospitais. *Revista Brasileira de Análises Clínicas, 45*(1–4), 7–13.

Pereira, J. C. M (2003). *Medicina, saúde e sociedade.* Ribeirão Preto: Complexo Gráfico Villimpress.

Rego, S. T. A. (1993). A medicalização do hospital no brasil: Notas de estudo. *Revista Médica de Minas Gerais, 3*(1), 54–57.

Santos, M. (1996). *A Natureza do espaço. Técnica e tempo. Razão e emoção.* São Paulo: Hucitec.

Santos, M. (2000). *Por uma outra globalização: Do pensamento único à consciência universal.* Rio de Janeiro: Record

Santos, M. (2008 [1994]). *Técnica espaço e tempo: Globalização e meio técnico-científico-informacional.* São Paulo: Edusp.

Schraiber, L. B. (1993). *O Médico e seu trabalho: Limites da liberdade.* São Paulo: Hucitec.

Sevcenko, N. (2009). O prelúdio republicano, astúcia da ordem e ilusões do progresso. In N. Sevcenko (Ed.), *História privada no Brasil 3. República: Da belle époque à era do rádio* (pp. 7–49). São Paulo: Companhia das Letras.

Teixeira, L. A. (2010). O Controle do câncer no Brasil na primeira metade do século XX. *História, Ciências, Saúde Manguinhos, 17*(1), 13–31.

Tubino, P., & Alves, E. (2009). História da cirurgia. Retrieved April 2, 2016, from https://alinesilvalmeida.files.wordpress.com/2010/05/historia_da_cirurgia.pdf.

Vianna, C. M. M. (2002). Estruturas do sistema de saúde: Do complexo médico-industrial ao médico-financeiro. *Physis, 12*(2), 375–390.

Chapter 6
The Political Economy of Territory and Agribusiness in Brazil

Samuel Frederico and Marina Castro de Almeida

Abstract This chapter aims to demonstrate how the notion of political economy of territory, proposed by Milton Santos, aids in the interpretation of spatial dynamics of Brazilian agribusiness at the beginning of the 21st century. Among the many territorial expressions of agricultural dynamics, this article analyzes the dialectical relationship between the 'centrifugal forces', exemplified by the spatial dispersion of modern agriculture, and the 'centripetal forces', represented by the centralization of production control especially in the metropolis of São Paulo. Since the Brazilian exchange crisis of 1999, state policy to stimulate export of agricultural products, linked with the interests of the main representatives of agribusiness (large producers and corporations), has resulted in the accelerating expansion of the agricultural frontier, especially with soybean production in savanna areas. However, concomitant with the territorial dispersion of production, the centralization of capital and the increasing influence of finance and information for the development of modern agriculture have reinforced the command function of the main national metropolis.

6.1 Introduction

Few ideas have had as much importance in the work of Milton Santos as that of the political economy of territory (Santos 1977, 1979, 1994, 2001). His proposal to consider the active role of geographical space in the formulation of the political economy is present in his work from the beginning of the 1970s up to his last book *O Brasil: Território e Sociedade no Início do Século XXI*, written with Maria Laura Silveira in 2001. Our point of departure in this present chapter is that the Brazilian state's adoption of policies to stimulate the exportation of primary products (mainly

Samuel Frederico, Assistant Professor, Department of Geography, Universidade Estadual Paulista—UNESP, Brazil; Email: sfrederico@rc.unesp.br.

Marina Almeida, Assistant Professor, Department of Geography, Federal University of Triângulo Mineiro—UFTM, Brazil; Email: marinacastrodealmeida@gmail.com.

© Springer International Publishing AG 2017
L. Melgaço and C. Prouse (eds.), *Milton Santos: A Pioneer in Critical Geography from the Global South*, Pioneers in Arts, Humanities, Science, Engineering, Practice 11, DOI 10.1007/978-3-319-53826-6_6

soybeans), together with the spread of information technology (as telematics) from the 2000s, led to a political economy founded on agribusiness—a process in which geographical space has been central.

Since then the Brazilian territory has been organized and used[1] as the object of various strategies and political maneuverings by the main agents involved in globalized agribusiness: the state, large producers, landowners, agricultural companies and financial investors. The acceleration of the opening of new productive regions, the implanting of fixed capital assets (such as highways, railways, ports, cities and communications systems) and the constitution of new territorial divisions of labour have spread new social relationships and spatial arrangements across the Brazilian territory. These new relationships and arrangements are almost always associated with the selective appropriation of surplus capital in renewed form.

We believe that Santos's notion of the political economy of territory offers a way to interpret the productive re-structuring and the Brazilian territorial dynamic linked to agribusiness at the beginning of the 21st century. Among the various ways of demonstrating how this new political economy of territory is structured, this chapter seeks to analyze, albeit briefly, the articulation between *centrifugal* and *centripetal forces* (Santos & Silveira 2001). To this end, we analyze the relationship between the expansion of the modern agricultural frontier (centrifugal forces) and the centralization of the productive command (centripetal forces) specifically in the metropolis of São Paulo.

6.2 Toward a Political Economy of Territory

Among the various meanings historically attributed to the term political economy, according to Santos (1994) the notion refers mainly to the study of relations of production. As the author emphasizes, scholars in this vein accentuate that the conditions of fulfillment and distribution of economic surplus arise, in the last analysis, from the complex relationships between capital and labour, owners and non-owners. However, the author himself (Santos 1979, 1994) reminds us that it is impossible to think of political economy without taking space into consideration.

Our purpose is to understand how the manifestation and the consequences of a particular process of production—in this case, agribusiness—organize the respective territory. And, dialectically, how the natural and socially produced conditions of the territory also shape production. Within this perspective, Santos (1994) defines geographical space as the historical result of the interaction between dead labour—accumulated in the forms of infrastructure and machinery—and living labour.

---

[1]For Santos (1996), the social object of analysis is not the territory itself, but the 'used territory'. This notion refers to the implantation and the differentiated use of engineering infrastructure or systems that arise from the dynamic movement of the economy and of society. To learn more about the concept, see Silveira (2014).

Santos articulates the political economy of territory as a set of theoretical propositions that includes space, network, region, used territory and socio-spatial formation. The notion has as its objective the overcoming of 'pure concepts' inherited from modernity and characterized by dichotomy (society/nature, technique/politics, object/action), by hybrid notions characterized by the indissolubility between natural and social, materiality and action. The objective of social analyses, then, is to explain space in terms of its use and not simply as an inert material area. From space's hybrid character we may discern a political economy of territory, that is to say, understand how production is undertaken in light of its power relationships (both of capital and labour) while recognizing the active role of space.

In attempting to operationalize the notion of the political economy of territory, Santos and Silveira (2001) propose interrelated analytical categories that include: the territorial divisions of labour; the dialectic between centrifugal (dispersive) and centripetal (agglomerative) forces; productive specialization and the shaping of new productive regions; the centralization of capital and spatial hierarchy; the corporative and competitive use of space; and 'unnecessary circulation'.[2] These are categories that frequently condition and result from one another.

In summary fashion that necessarily runs the risk of theoretical imprecision we can affirm, inspired by Santos and Silveira (2001): (1) that the concentration and centralization of capital in a few hands and locations favors the expansion of the territorial division of labour of companies and productive activities; (2) that the incorporation of subspaces into the logic of dominant production leads to territorial productive specializations which are 'alien', because they are imposed from outside, and 'alienated', because 'they are neither locally inspired nor concerned with local destiny' (p. 300); (3) that new productive regions are created that, generally subordinated to world standards of competition (quality, productivity and costs), specialize in particular products, especially for exportation, and are characterized by their technical competence (means of production), while the political aspects of this process (be they normative, financial and informational) are decided externally; (4) that the establishment of a relation between the 'centrifugal forces'—the expansion of the modern productive area—and the 'centripetal' ones—the concentration of the decision-making poles—reinforces a 'spatial hierarchy'; and (5) that the more the places concerned specialize, the greater becomes the need to transport their products over great distances—especially when intended for exportation—which ends up generating 'unnecessary circulation', that is to say, the construction of practically mono-functional systems of transport, because they are dedicated to the exportation of a few products, which instead of generating local benefits becomes being an excessive onus on society as a whole, especially in

---

[2]With the idea of unnecessary circulation, Santos and Silveira (2001) are theorizing in parallel with the Marxist idea of necessary and unnecessary production, i.e. 'a production whose presence is able to ensure the well-being of the population in comparison with other production intended export' (p. 297). From the moral point of view, necessary production and circulation would be those that help people to survive and to develop, while unnecessary production and circulation would be 'excessive, leaving to society an unnecessary burden' (p. 297).

countries with great social and regional inequalities and lacking in social infrastructure such as Brazil.

The expansion of modern agriculture in Brazil is one of the best examples of the aforementioned overlapping territorial dynamic. Within agricultural areas there is a veritable rearrangement of territory by the demands of globalized agribusiness (Elias 2011). On the one hand, the relations of power between the main agents of agribusiness (the agro-industry, producers, landowners and the state) play a central role in territorial organization leading to the installation of infrastructure, the construction of cities, migration of the population, the labour market, the landownership structure and the characteristics of production. On the other hand, in indistinguishable, dialectical fashion, the continental dimensions of Brazilian territory, the existence of areas favorable to the expansion of modern agriculture, the favorable physiographical conditions (soil, rainfall, water reserves and relief) and the selective presence of infrastructure (highways, railroads, waterways, warehouses and energy) also determine agricultural quality, costs, localization and production (Frederico 2010; Castillo & Frederico 2010).

6.3 Pact of the Political Economy of Territory Based on Agribusiness

In the last decade of the 20th century the subordination of the Brazilian state to a financial and neoliberal logic through the implementation of the Washington Consensus led to the strengthening of exportation policy, which was intensified after the exchange crisis of 1999 (Delgado 2012). This is what Santos (1999[3]) refers to as the 'imperative of exportation', that is, the creation of policies with the deliberate purpose of increasing Brazilian exports, with a view to generating external trade to pay foreign debt and to cover the current account deficit, thus inverting the resulting reduction in Brazil's international reserves.

It was for this purpose that the agribusiness sector was again selected by the state, as had been done at other moments in Brazil's economic formation,[4] to be the main actor responsible for saving the country's external macroeconomic policy. This led to significant alterations in the organization and use of territory in the areas of modern agriculture, such as: the accelerating expansion of the agricultural frontier, replacing native vegetation and small-scale family agriculture; the intensification of production and of regional productive specialization; the growth in the concentration of land-ownership; and the creation of a new territorial division of labour of large corporations and of the activities of agribusiness, alongside the

---

[3]Santos, Milton 1999: 'Guerra dos Lugares', in: *Folha de S. Paulo. Caderno Mais!*, 8 August.
[4]Such as the incentive policies to export coffee in the 1960s, in order to generate foreign exchange for the implementation of industrialization policies.

construction of practically mono-functional infrastructure for the purpose of facilitating the outflow of produce (Bernardes 2006; Frederico 2011; Elias 2011).

The materialization of these events within Brazilian territory is due to the fact that the political economy of territory is now based on agribusiness. That is to say that the incorporation and re-organization of space has been accomplished to satisfy the interests and the accumulation strategies of the aforementioned agents of globalized agribusiness.

Due to the limited scope of this chapter and of the innumerable possibilities of demonstrating this pact among agribusiness that comprises the political economy of territory, we have decided to focus our analysis on the dialectic between the centrifugal and centripetal forces involved. One is here dealing with centrifugal forces of dispersion—with the advance of the agricultural frontier that has gathered speed since the 2000s. These centrifugal forces have been articulated with centripetal forces—with the strengthening of the decision-making power of some locations within the territory, with the centralization of capital, and with the valuation of finance and information as productive activity. We elaborate on these forces in the following section.

6.4 Centrifugal and Centripetal Forces: The Expansion of the Agricultural Frontier and the Commanding Role of the Metropolis of São Paulo

By the term 'modern agricultural frontier' we mean those areas of the Brazilian territory occupied from the 1970s by capital and technology-intensive monocultures that have replaced areas of original vegetation (principally of *Cerrado*)[5], of subsistence agriculture (practiced by peasants) and of extensive pastures (Frederico 2010).

Motivated by economic and geopolitical factors, the centrifugal expansive force of the agricultural frontier occurred primarily in the direction of those *Cerrado* areas, which correspond to a quarter of the total extent of Brazilian territory (i.e. about two million km^2). Although it has been continuous, the speed of expansion of the agricultural frontier occurred in three distinct moments during the period from 1970 to 2010. There were two phases of rapid expansion during the periods 1970–1985 and 2000–2010, mainly stimulated by the government's policy that sought to increase the exportation of primary products and encouraged by a generous offer of subsidized rural credit. The frontier continued to expand, though at a slower pace, between 1985 and 2000. In this intermediate phase, characterized by the exchange and fiscal crisis of the Brazilian state (in the 1980s) and by the adoption of neoliberal ideology and policies (in the 1990s), there was a significant reduction in

---

[5]The *Cerrado* is a Brazilian biome of the savanna type located principally in the central region of the country.

the offer of public credit for agricultural activity, with a decrease in the rate of expansion of the frontier.

Especially from the 2000s the establishment of the 'pact of the political economy of agribusiness' led to a reaccelerating expansion of the agricultural frontier. According to data of the *Instituto Brasileiro de Geografia e Estatística* (Brazilian Institute of Geography and Statistics) (2014), the area occupied by temporary crops on the agricultural frontier attained 24.2 million hectares in 2012, as compared with 11.6 million in 2000, and the area dedicated to soya production (the main Brazilian export crop) more than doubled in the same period, from 6.4 to 13.8 million hectares. As a result of this expansion, the region encompassed by the modern agricultural frontier became the greatest Brazilian producer of cotton (96.5%), sunflower (90.5%), sorghum (71.8%), soya (64.1%) and maize (48%) and the second largest of sugar-cane (16.3%) and rice (10.7%) (PAM/IBGE 2014).[6]

With this expansion, there has also been a growing productive territorial specialization, that is to say, the 'increase within one and the same region of the diversification of the tasks related to one single process while the other techniques and forms of work diminish' (Silveira 2011: 79). In 1990, of the 50 largest Brazilian municipalities that produced soya, 25 were not a part of the aforementioned agricultural frontier. In 2012 this number had been reduced to just three (two in the state of Paraná and one in Minas Gerais). In that year, among the almost 1,800 Brazilian municipalities that produced soya, only 76 were responsible for 50% of the total volume, while the top 10 produced 16%.

The expansion and productive specialization in frontier areas is even more evident when we consider the expansion of the temporary crops (cultures that need to be replanted every year) of the ten greatest producers of soya[7] (all situated on the agricultural frontier). In the municipalities that host these producers, the planted area increased by an average of 72% in the 1990s and by 200% in the 2000s. In all of them, the total area planted with soya, maize and cotton accounted for more than 90% of the total cultivated area in 2012 (PAM/IBGE 2014).

Dialectically, the territorial expansion of modern agriculture has seen the simultaneous centralization of its command, which has also intensified from the 2000s. This is due to the increasing presence in modern agricultural production of major national and international corporations whose head-offices are usually located in the metropolis of São Paulo. While the technical control of the productive operations is relatively dispersed, with the notable presence of engineers, consultants and agronomists in the agricultural regions, the normative, financial and informational regulation tends to be concentrated in a select number of locations such as the metropolis of São Paulo. It may be stated, even if only tentatively, that

---

[6]The percentages refer to the total of Brazilian production. PAM/IBGE. 2014: Instituto Brasileiro de Geografia e Estatística. 'Produção Agrícola Municipal'. Série Histórica 1990–2012. Retrieved August 18, 2016, from http://www.sidra.ibge.gov.br/bda/pesquisas/pam/default.asp.

[7]In decreasing order: Sorriso (MT), Sapezal (MT), Nova Mutum (MT), Campo Novo dos Parecis (MT), Formosa do Rio Preto (BA), Rio Verde (GO), Nova Ubiratã (MT), Querência (MT), Diamantino (MT) and Jataí (GO).

the metropolis of São Paulo is ubiquitous not only on the agricultural frontier but also in the majority of the Brazilian regions dedicated to modern agriculture, as well as having an increasing influence in other South American countries.[8]

The growing centralization of agribusiness capital contributes to the reduction of the number of decision-making centres. When national companies grow, they prefer to transfer their head offices to the main metropolises, especially São Paulo. In the cases of international companies or of national ones incorporated by international capital, the capital city of São Paulo state ends up exercising a *delegated* form of control—an intermediate stage between the corporative head offices in other countries and the Brazilian productive regions. Maintaining the office in São Paulo is of strategic importance to the company, because the location makes convenient the finding of qualified staff, and makes easier the undertaking of face-to-face business between the company's directors, consultants, bank managers and directors of consultancy firms.

Of the 50 largest companies involved in the Brazilian agribusiness in 2011, 25 had their head-offices in São Paulo, as did the four largest international agricultural trading companies (Cargill, ADM, Louis Dreyfus and Bunge), as well as the main companies of sugar and ethanol (Raízen and Copersucar), wood and cellulose (Suzano, Klabin, Fibria and Duratex), meat (JBS and Marfrig), and fertilizer and agrochemical (Basf, Bunge Fertilizantes, Bayer, Syngenta and Mosaic).

The municipality of São Paulo is the largest Brazilian exporter of soya, sugar, maize, ethanol and meat, as the offices of the companies responsible for the foreign operations of purchase and sale are situated there. In 2013 the soya complex (beans and derivatives) together with the sugar and energy sector (sugar and ethanol) accounted for 50% of the total value of the exports of the municipality. While the municipality of Sorriso (MT)—the largest Brazilian producer of soya with a crop of about two million tons in 2012—exported 1.47 million tons (beans and derivatives) for a total value of US$ 814 million, São Paulo, without having produced a single soya bean of its own, exported 2.97 million tons with an approximate value of US$ 1.6 billion. This was a little more than 5% of the quantity and the value of all the national exports of the complex in 2013. A similar trend has occurred with all of the other agricultural and livestock breeding commodities. For example, sugar-cane products accounted for almost 30% of São Paulo's exports, with 5.23 million tons valued at US$ 2.44 billion in that same year (SECEX/MDIC 2014).[9]

The centralization of the vectors of command creates a new political economy of territory with the strengthening of spatial hierarchies and a greater rigidity of relationships between the polarizing and the polarized areas. Although the majority of agribusiness companies have many small offices scattered over the main agricultural regions, the major deals (involving, for instance, exportation, importation

---

[8]Ubiquity means the power of the head-offices of large companies located in São Paulo to define the organizational and productive aspects of agricultural regions.

[9]SECEX/MDIC. 2014: Secretaria do Comércio Exterior. Ministério do Desenvolvimento, Indústria e Comércio Exterior. Estatísticas de Comércio Exterior. 2010. Retrieved August 18, 2016, from http://www.mdic.gov.br.

and loans) and strategic decisions (such as land acquisitions, agreements with companies and logistics) continue to be centralized in the offices in the city of São Paulo (Frederico 2015).

The city of São Paulo is also beginning to attract foreign South American companies as is the case of the Argentinian firms Adecoagro and El Tejar. The former is one of the most apt examples to illustrate the territorial division of labour among the productive regions and the metropolis of São Paulo. Employees specialized in agricultural operation locate in cities close to farms, such as in the region west of Bahia. These employees include agricultural, administrative and financial managers of the farms, the coordinators of machines and operations (planting/harvesting), and the agronomists and agricultural technicians. In agribusiness head offices in São Paulo—situated in the suburb of Moema, one of the most prosperous regions of the city—are the general and agricultural directors, the control centre of the company, the legal department, the area of strategic planning and the commercial and financial team.

The centrality of commodity production and the intricate relations between agricultural and financial capital have also reinforced the commanding role of the city of São Paulo (Santos 1994). Since the end of the 19th century the main stock exchanges of the world (e.g. New York, Chicago and London) have been involved in regulating the international prices of agricultural commodities. This has happened in Brazil too—through the Santos Exchange for coffee. However, with the adoption of neoliberal policies and the so-called 'deregulation' of the world economy (Chesnais 2005), especially from the 1980s, the volume of commodity business on the merchandise and futures exchanges has increased significantly (Clapp 2009). In Brazil it has been no different: the creation of BM&FBovespa in 2008, by the fusion of the São Paulo Stock Exchange (Bovespa) and the Bolsa de Mercadorias e Futuros (Merchandise and Futures Exchange) (BM&F), consolidated the transactions of various other Brazilian exchanges. At present BM&FBovespa is active in the futures market of the main Brazilian agricultural commodities soya, coffee, maize, ethanol and cattle. The majority of the business monopolies created over the past decade were built through agreements among large national companies and with foreign capital (Brasil Foods, SLC Agrícola, Brasil Agro, Heringer, Agrenco, Laep, JBS, Marfrig, Minerva, São Martinho, Açúcar Guarani, Cosan, Tereos, Suzano, Fibria, Klabin, Duratex, Eucatex and Ecodiesel)—their shares are now traded on the BM&FBovespa.

In financial terms, São Paulo was the Brazilian city that withdrew the largest number of rural credit loans in 2012. Between 2000 and 2012, the volume of rural credit increased from R\$ 150 million to almost R\$ 1.4 billion, approximately 10% of the national level (BCB 2015).[10] Despite the municipality's ostensibly negligible agricultural and livestock production, the large volume of loans can be explained by the presence of the largest national banks and international suppliers of agricultural

---

[10]BCB (Banco Central do Brasil), 2009: Anuário Estatístico do Crédito Rural, Série Histórica 2000/2012. Retrieved August 18, 2016, from http://www.bcb.gov.br/?RELRURAL.

credit, such as Banco do Brasil, Bradesco, Itaú, Rabobank, HSBC, Votorantim, ABC Brasil, Citybank and Santander. The physical presence of these agencies in the city matters because, very often, the largest loans require the personal participation of the loan-requesting and loan-granting CEOs. The main clauses of relevant loan contracts are negotiated at the business meetings and lunches in which the directors of both concerned institutions participate. It is at these face-to-face meetings, in the presence of their respective legal teams, where the guarantees demanded by some banks are defined, such as the provision of strategic information and participation in future business of the agricultural company. Some large producers—borrowers of considerable amounts of credit—also travel to São Paulo to obtain the loans personally from the managers of the largest bank agencies.

On the one hand, the quest for greater productivity and profitability, in accord with the demands of the globalized market, leads to productive specialization. This results in the constitution of sub-spaces in Brazilian territory in which monocultures for export predominate on large properties at the expense of other crops and forms of production. On the other hand, the growing dependence on modern agricultural activity—in terms of finance and information of a strategic, legal, marketing, accounting and logistic character—makes the role played by the metropolis of São Paulo even more significant in the regulation of production. Thus a political economy of territory is established, characterized by a filigree of territorial divisions of labour, that is, by the distribution and connectivity of the places and regions of interest for corporate agribusiness activity.

6.5 Final Considerations

As we have tried to demonstrate in brief fashion in this text, the idea of the political economy of territory as proposed by Milton Santos provides a significant theoretical-methodological means of interpreting the productive transformations and Brazilian territorial dynamics at the beginning of the 21st century. The Brazilian state's policy for stimulating the export of agricultural commodities, together with the interest of the principal agents of globalized agribusiness, have led to the creation of a political economy of territory based on agribusiness. We are here dealing with the organization and use of territory for the main purpose of making possible the increase of agricultural and livestock production by means of the expansion of the modern agricultural frontier, and the intensification of production.

Among the various examples that demonstrate this pact of the political economy of territory, we chose the articulated and dialectical movement of the centrifugal and centripetal forces (respectively of dispersion and concentration). As Santos (1994) proposes, with the gradual diffusion of the networks of communications and transport, especially from the 1970s, a unitary logic was created within Brazilian territory. This logic strengthened the combined movement of the territorial dispersal of modern activities (agriculture, industry and services) and the concentration of the political command, especially in the metropolis of São Paulo.

At the beginning of the 21st century state policies to stimulate the territorial expansion of modern agriculture, coupled with the possibilities offered by information technology, reinforced the logic of expansion and concentration. Parallel to the acceleration of the dispersal of modern activities, there is an increase in the concentration of the political command of the productive processes. Even if the technical control of the productive operations is relatively scattered, its normative, financial and informational regulation is becoming increasingly centralized in Sao Paulo's metropolis.

There is, thus, a growing territorial division of labour among areas specialized in technical operations and areas dedicated to political command. This process creates a growing subordination of the local order to corporative global interests and dictates. As a consequence, in the logic of the state and of the large agribusiness companies, the new pact of the economic policy of the territory represents solutions, even though temporary, for macroeconomic problems (such as the exchange crisis) and opportunities for investment. However, on the ground, the new forms of use and organization of territory almost always promote conflict and greater socio-spatial segregation once the infrastructure is built and used selectively, aiming almost exclusively at the export of few primary products. This spatial differentiation can be considered one of the derivations of the establishment of a political economy of territory based on agribusiness at the beginning of the 21st century in Brazil.

References

Bernardes, J. A. (2006). Circuitos espaciais da produção na fronteira agrícola moderna: BR-163 matogrossense. In J. A. Bernardes & O. L. Freire Filho (Eds.), *Geografias da soja: BR-163 fronteiras em mutação*. Rio de Janeiro: Edições Arquimedes.

Castillo, R., & Frederico, S. (2010). Dinâmica regional e globalização: Espaços competitivos agrícolas no território brasileiro. *Mercator, 09*, 17–26.

Chesnais, F. (2005). O capital portador de juros: Acumulação, internacionalização, efeitos econômicos e políticos. In F. Chesnais (Ed.), *A finança mundializada*. São Paulo: Boitempo.

Clapp, J. (2009). Food price volatility and vulnerability in the Global South: Considering the global economic contexto. *Third World Quarterly, 30*(6), 1183–1196.

Delgado, G. C. (2012). *Do capital financeiro na agricultura à economia do agronegócio: Mudanças cíclicas em meio século, 1965–2012*. Porto Alegre: UFRGS Editora.

Elias, D. (2011). Agronegócio e novas regionalizações no Brasil. *Revista Brasileira de Estudos Urbanos e Regionais (ANPUR), 13*, 153–170.

Frederico, S. (2010). *O Novo tempo do cerrado: Expansão dos fronts agrícolas e controle do sistema de armazenamento de grãos*. São Paulo: AnnaBlume.

Frederico, S. (2011). The modern agricultural frontier and logistics: The importance of the soybean and grain storage system in Brazil. *Terrae, 8*, 24–32.

Frederico, S. (2015). Economia política do território e as forças de dispersão e concentração no agronegócio brasileiro. *GEOgraphia, 17*(35), 68–94.

Santos, M. (1977). Sociedade e espaço: A formação social como teoria e como método. *Boletim Paulista de Geografia, 54*, 81–99.

Santos, M. (1979). *Economia espacial: Críticas e alternativas*. São Paulo: Hucitec.

Santos, M. (1994). *Por uma economia política da cidade*. São Paulo: Hucitec/Educ.

Santos, M., & Silveira, M. L. (2001). *O Brasil. Território e sociedade no início do século XXI*. São Paulo: Record.

Silveira, M. L. (2011). Território usado: Dinâmicas de especialização, dinâmicas de diversidade. *Ciência Geográfica, 15*(1), 04–12.

Silveira, M. L. (2014). Used territory: A kaleidoscope of spatial division of labor. *Revista Geográfica del Sur, 5*(7), 15–34.

Chapter 7
Territorial Planning in Brazil: An Interpretation Based on the Ideas of Milton Santos

Luís Angelo dos S. Aracri

Abstract This text discusses how Milton Santos's work provides elements to understand two major distinct phases of territorial planning policies in Brazil. The first major phase, which lasted approximately half a century, until the late 1980s, is analyzed from the deconstruction of the theoretical foundations provided by the so-called 'regional science'; the second, extending from the 1990s to the present day, is interpreted based on the concepts Santos has developed for understanding globalization. Despite the differences in the logic of these two phases, there are continuities of traits.

7.1 Introduction

Despite today's widespread regionalization of production, international trade, and global competitive regimes, national-scale planning has not decreased in importance. On the contrary, state-planned interventions in national territory are as relevant today as in previous periods. Of course there have been some transformations, and it is necessary to understand how these changes have taken place. In this short essay I aim to provide readers with an overview of territorial planning policies in Brazil, from their inception to today. I interpret the past and present dynamics of planning through key theorizations of Brazilian geographer Milton Santos.

It is impossible to exhaustively detail so vast a subject in the following pages, thus in this text I focus on a few ways of interpreting the role played by territorial planning in Brazil. The prolific Santos provides many analytic perspectives for understanding this phenomenon. For necessity of brevity, the analysis I present here is deliberately partial and humbly introductory.

Luís Angelo dos S. Aracri, Associate Professor, Department of Geosciences, Federal University of Juiz de Fora, Brazil; Email: luis.aracri@ufjf.edu.br.

© Springer International Publishing AG 2017   79
L. Melgaço and C. Prouse (eds.), *Milton Santos: A Pioneer in Critical Geography from the Global South*, Pioneers in Arts, Humanities, Science, Engineering, Practice 11, DOI 10.1007/978-3-319-53826-6_7

I discuss the development and the implementation of territorial planning policies in Brazil in two sections. The first section focuses on a period which officially started in the 1930s, when planning was used as an instrument in the projects of geopolitical and market integration of the national territory; diversification and expansion of industrialization; and the general dynamics of postwar capitalism. The events taking place between the 1950s and 1980s constitute my main focus during this period. The second section deals with the period that started in the 1990s and continues until now. It is marked by a shift in the focus of territorial planning toward an autonomous, more competitive insertion into the so-called 'global market'. This insertion is done through the preparation and qualification of subnational territories, that is to say, through investments in new infrastructures and modernization of those already in existence, but also through strategies and policies to attract companies and private investors. The intention in this period is to attract public and private investments and eventually connect national and international circuits of goods and information.

Two of Santos's works are fundamental for understanding certain aspects of planning policies in these two periods. *Planning Underdevelopment and Poverty* (in Santos 2003) is particularly relevant to the first. This text offers scientific/theoretical foundations for the territorial planning policies that were in effect from the 1930s to the 80s. This work points to some of the effects of those policies on Brazilian territory, economy and society as a whole. With reference to the latter contemporary period, I draw from one of Santos's classic works, *A Natureza do Espaço* (*The Nature of Space*). In this book Santos (2002) elaborates theoretical concepts such as *war of places*, *spatial productivity*, *national spaces of international economy*, *technical networks* and *fluidity*. These concepts can help us to better understand the current effects of planning on the territorial (re)ordering of the Brazilian economy.

Santos's first critical studies about regional and urban planning in the then-called 'Third World' were released in the 1970s. Several of his notions and expressions are certainly dated. Many are still relevant, however, and it is necessary to reinterpret them, to redefine them, to acquire a new content for them. The debate must continue.

7.2 The 'First Era' of Territorial Planning in Brazil

In Brazil, territorial planning was established by state agencies as an institutional instrument whose purpose was to adapt productive activities, production factors and social relations to the economic models implemented in the country. These economic models originated in the most advanced capitalist economies, as pointed out by Santana (2011). According to Santana, for five decades, from 1930 to 1985, the state put into practice projects and plans supported by integrationist and developmental theories, created regional planning bodies and, in an openly geopolitical position, proclaimed territorial occupation as a necessity to increase the use of natural resources and defend its borders.

Territorial planning[1] in Brazil during this period enabled a fast and effective engagement of big business with the national economy. It also (re)created social and spatial inequalities in the country (Santos 2003), despite being couched in a discourse of overcoming poverty and regional disparities. According to its advocates, planning would be a 'tool' used to achieve development, which was seen, at the time, as a synonym for economic growth and for the adoption of the same consumption patterns as found in developed countries.

In order for Brazil and other peripheral countries to achieve economic growth it was considered necessary to expand investments, mainly through major infrastructure projects. Nevertheless, the investments were very costly and these countries were forced to take out loans or to encourage the incorporation of private international capital into their economies. These measures resulted in export-oriented agricultural production, the exploitation of mineral resources, the increase of dependence on foreign countries, and domestic and foreign indebtedness.

Araújo's (2000) analysis describes these processes with respect to Brazil. Between the 1950s and the 1980s, Brazil had as its main project the construction of a national industrial park. The major purpose of this project was to become a great world power as a new industrial country and to incorporate Brazil's economy into the overall dynamics of post-World War II capitalism. The national goals of economic growth and world economy integration coincided with the interests of productive and financial international capital. The latter, in turn, either established branches in the country or funded the industrial park project. The economic history of Brazil in this period thus began to overlap with the history of the national industrializing project, which was marked by the fast formation and consolidation of the 'domestic market'. As a result, the country's economy gradually shifted from local-based to national-based activities.

There were a number of processes engendered by this relationship between the modernizing/industrializing project and territorial planning in Brazil during the period (Araújo 2000: 18–20): (i) fomentation of the connections between the metropolis of São Paulo state and other regions in the country (1960s–1970s); (ii) productive integration based on the regionalization of large oligopolistic industry, strengthened by commercial exchanges in different parts of the territory (1970s); (iii) surplus production in the trade balance so as to obtain currencies to pay the foreign debt (1960s and 1970s); and (iv) physical-territorial integration with the establishment of extensive transportation and communication facilities.

Territorial planning intervened in each of these processes through different strategies: fiscal and financial incentives for the transfer of productive capital into less industrialized areas; further incentives for the use of land and extraction of natural resources considered abundant; and the creation of 'management plans' for the implementation of infrastructures, with a large emphasis on developing the road network.

---

[1]Territorial planning is considered in this text as a coordinated set of policies for the spatial redistribution of investments, resources, public services, infrastructure and revenues. It is a tool of ordering and management of the territorial distribution of economic and demographic growth, and (in principle) social welfare and environmental preservation.

Territorial planning, as planned intervention by the state, from the post-World War II period until the 1980s was articulated with so-called 'regional science'. This science, according to Santos (2003: 19), is a 'spatial theory at the service of capital'. The development of regional science began after World War II, as economists and geographers focused on economic and spatial problems, respectively. One of the central concerns of regional science was how to attract and concentrate resources and capital in particular regions. In other words, it tried to understand why some spaces tend to achieve higher economic development than others. Regional science theorists, along with their supporters in government, did not believe that market forces alone were capable of inducing economic development, and they thus supported state interference in the economy through territorial planning.

According to Santos, the pairing of regional analysis with territorial planning was based on theories such as Walter Christaller's 'central places', which explained the existence of large concentrations of capital. Key actors in regional science and planning later attempted to couple central place theory with François Perroux's 'growth and development poles' theory (Lima 2003). Perroux's theory was based on the idea of a polarized economic space, that is, an abstract space formed by purely economic relationships among individuals, families, companies and the state, and focused on the connections between economic areas (e.g. between zones which housed certain industries and zones inhabited by the industries' respective supplier and consumer markets). Spatial interaction here is simply a relationship of interchange or interdependence (flows) between homogeneous spaces called poles. Poles generate forces of attraction and repulsion and each pole has its own field of action, thus forming polarized areas or spaces. Development poles would be those leading to structural changes and encompassing the entire region's population.

According to Egler (1995), the theory of growth and development poles was very much employed by territorial planning in Brazil. The state specifically used this theory in developing domestic markets and establishing conditions for linkages with foreign markets. The state fostered the formation of poles at selected points in the national territory, and the connection between them formed market and financial circuits, thus promoting market integration. These privileged points were also the best able to connect with other foreign financial and goods circuits.

Santos (2003) observed that Christaller's emphasis on large spatial concentrations helped territorial planners enact Perroux's pole theory. It also gave new strength to the 'innovation diffusion theory' promoted by Swedish geographer Torsten Hägerstrand. Spatial diffusion of innovation refers to the propagation, in space, of modern productive activities (such as large industry). This theory classifies spaces as output innovation centres (a city or a pole area) and input centres (peripheral areas); the diffusion process results from the interactions between these output and input centres, between developed and less developed regions. Following this approach, the purpose of planning is to create conditions that favor these interactions, thus promoting the diffusion of economic development from one region to another.

Between the 1950s and 1980s Brazil evidenced not only the commitment and loyalty of national planning bodies to these theories, but also exposed their problems. The case of Brazil's Northeast region is a case-in-point. The state constructed communication and transportation networks in order to integrate this major region with the core area of the country (the Southeast), and created the planning body SUDENE (Superintendency for the Development of the Northeast) to execute large infrastructural projects. However, nothing could break the dependence of the Northeast on the Brazilian Central-South (Goiás, Mato Grosso and Mato Grosso do Sul). One of the gaps in diffusion innovation theory is understanding how power asymmetries function; territorial planning attempted to create poles of development, but unfortunately could not overcome regional elites' monopoly on rural property.

As a result of this and other failed development initiatives, Santos concluded that all territorial planning would have the same effect regardless of its theoretical underpinnings: no matter how large the investments in 'secondary centres' or even in 'underdeveloped regions', the returns would always be routed to big centres. He argued that adjustments of economic activities, production factors and social relations of production in Brazil and in the 'underdeveloped world' more generally would have very fast and brutal effects.

The forces of transformation thus did not proliferate across national space homogeneously; on the contrary, the impact of those forces was very local and selective, and eventually worsened the existing imbalances. Santos (2004) drew attention to the importance of considering territorial specificities in countries belonging to the once-called Third World in light of these development projects. In these countries: (i) there are enormous social income differences expressed at the regional scale, manifest in a strict hierarchy of activities, and at the local scale, manifest by the coexistence of activities of similar nature but aimed at different consumption patterns; (ii) space is subject to and affected by multiple decision-making levels so that the smaller the place the more diverse the impacts generated; and (iii) only some parts of the territory welcomed economic modernization projects, and those that did, did so at different moments and in diverse ways. To Santos, capitalist modernization in Brazil throughout the 20th century, facilitated by territorial planning, was profoundly shaped by a selective and unequal spatial history—resulting in 'planned poverty' (Santos 2003: 29), in his own words.

7.3 The Current Dynamics of Territorial Planning in Brazil

The oil crisis of the 1970s affected new forms of accumulation in Brazil in the following decade (Araújo 2000). If, before the crisis, petroleum was the main raw material to be outsourced, the new accumulation regime that emerged in the wake of the crisis was premised on knowledge/information as primary resources. With the progress of the so-called 'technological-scientific revolution' (Santos 2004), new sectors of activity, which were embedded in 'central countries'—countries with higher levels of social and economic development—began commanding global economic dynamics. Meanwhile, Brazil was plunged into a crisis of domestic and foreign indebtedness; the transfer of savings abroad for debt-repayment to international creditors had stopped national investments. The industrializing model of previous decades appeared to have come to an end.

Between 1985 and 1995 territorial planning policies shifted in the state's agenda, partly as a result of these global economic dynamics. Planning became focused more on sector-specific or strictly economic plans, rather than on the previous large, cross-cutting regional policies. Even so, regional planning remained important: the 1988 Constitution stated that the Executive Power should establish 'Pluriannual Plans' (PPAs) to define, by region, the directives, targets and objectives of federal public administration for the entire national territory.

The 1990s saw an increasing shift to neoliberal policies in Brazil, beginning with the victory of Fernando Collor de Mello in the presidential elections of 1989. According to Santana (2011), the first PPA presented by the Collor government, valid for four years (1991–1995), was less a planning policy than a budget plan. At the same time, the country was confronted with faster and deeper socioeconomic, political and institutional global changes. The increased mobility of foreign productive and financial capital forced the restructuring of Brazilian sectors that had not adapted to the new globalized forms of accumulation and deregulation of markets. As a consequence, there was increased private sector participation either in market intervention, through establishing prices and setting conditions (or barriers) for entry, or in political agreements, that is, agreements with workers and local powers with little-to-no state mediation.

In reaction to Collor's plan, territorial planning was resurrected during the first presidential term of Fernando Henrique Cardoso. Cardoso's Pluriannual Plan of 1996–1999 represented a paradigm shift. This time, the state played the role of a proponent of plans and policies rather than of a centralization agent in charge of the different planning steps. Additionally, the territory, formerly understood as an archipelago of semi-autonomous regional markets not much connected to each other—and which should be integrated in order to form a national market—began to be seen as an element that could connect national and international circuits of capital and information (Santos 2002). The plan still concerned territorial inequality, in theory, but the main focus of planning was to craft local spaces through a logic of territorial selectivity: cities and regions were to attract capital for the implementation of territorially-connected investments at different scales. This new approach to planning favoured the adoption of different instruments varying by region or city.

In general, the emergence of planning at the municipal level was a primary component of what Santos (2002) called the 'war of places', or the ways in which places competed amongst one another for investment. Santos argued that companies are not the only ones involved in global competition. Places are also in search of investments, of resources, of labour force and of consumers. The new information economy is central here: according to Santos, as information became one of the main forms of wealth accumulation, the *knowledge* of places globally allowed for more precise spatial selectivity than in the past. The actions devised in Cardoso's first PPA enabled federation units and states to invest, with maximum or minimum support from the government (depending on the arrangement of local and regional forces and the Federal Government), in certain productive sectors. War of places, or interterritorial competition, paved the way for Cardoso's second term PPA (2000–2003), which emphasized connections with the international market.

According to Santos (2002), spaces can be distinguished by, among other things, their differential capacity to generate profitable investment. The possibility of generating larger or smaller profits in a given sector or field of activity is related to how a given space is constructed, rather than a given space's 'natural characteristics'. The capacity of a place to create profit depends, above all, on local conditions of a *technical* nature—such as energy, transportation and telecommunications infrastructure—and also of an *organizational* nature—such as local laws, systems of fares and taxes, labour relations and traditions, and formal and informal institutions. In other words, urban and regional spaces/scales specialize in particular forms of production due to human-made technical, economic and political projects, which Santos referred to as *spatial competitiveness* and *productivity*. Specialization of places—urban or regional—meets the global competitive market's demands for security and profit. However, the productivity of a city or region may not be permanent, as other places compete to offer larger returns to those same production sectors.

The federal PPAs of Brazil in the late 1990s and early 2000s constructed different spaces for investment throughout the country. The territorial ordering of industry was accomplished in a number of ways, through: investments in modernizing and expanding roads, privatizing strategic sectors such as telecommunications (for decades under the monopoly of the state), and incentivizing the relocation of industries to less industrialized areas. The federal government also fostered aggressive competition between municipalities and states based on tax and fee exemptions, on donations of land, and on compensation to private companies for expenditure on infrastructure (Piquet 2007).

Industrial production was spread throughout the national territory. According to Lencioni (2006), the metropolitan area of São Paulo began to share the leadership of industrial production with the rural areas of the same state where new industrial districts had been created. In the state of São Paulo 'the industrial productive capital' moved inland, while the headquarters of large corporations, 'higher order services,' and research centres remained concentrated in the capital city. The state of Rio de Janeiro, despite tendencies toward de-industrialization and a rise in the trade and services sectors, has experienced a cycle of investments in petroleum, with oil companies opening their headquarters in the capital. The city of Rio has thus become a *technopolis* for the petro sector, while oil platforms and processing are concentrated in the Campos Basin.

There has also been a dispersion of industrial production to the Northeast region of the country. For example, Fortaleza recently built the largest industrial park in this region (Piquet 2007), taking the title from another Northeastern city, Recife.

The state of Ceará has also made concerted efforts to become the best, most time-efficient option for sea transportation. The irrigated fruit industry in the western part of Rio Grande do Norte, also of the Northeast region, has been an important source of wealth accumulation. Industry in this area represents a recovery from political and institutional barriers of historical oligarchies whose power was based on the monopoly of land and of the 'drought industry'. The development of the irrigated fruit industry in this region was greatly facilitated by the Armando Ribeiro Alves dam project, constructed in 1983 by the state government. The dam valorized the land in the Vale do Açu area, facilitated the introduction of irrigated crops, and attracted new rural owners who were more open to innovations in the science and technology sector. A local-based productive scale—with local production factors, infrastructures and political mediation—developed afterwards, but the international market scale remained as a preferential option.

The examples presented above illustrate how territorial planning has, through the Pluriannual Plans, contributed to redesigning the spatiality of the Brazilian economy. Industrial development has occurred under the aegis of a new project: strengthening territories' comparative advantages in order to attract and make profitable investments from specific globalized sectors. In other words, territorial planning has facilitated the autonomous and competitive insertion of different Brazilian regions and cities into the global economy. According to Santos (2002: 244) these territories have gradually turned into 'national spaces of international economy'. It is the international scale, after all, that defines the quality patterns of products, and transnational firms are the ones best able to make use of major infrastructure projects.

Territorial planning's emphasis on international trade also reflects broader regional concerns about, for instance, other Mercosul countries and Brazil's participation in the Initiative for the Integration of Regional Infrastructure in South America (IIRSA). In the second PPA of Cardoso's term, the Brazilian government consolidated partnerships with other South American nations in order to make the continent a preferential regional platform. As a result, so-called 'major investment projects', which played a key role in the first phase of territorial planning, have retained their importance, albeit in ways increasingly connected to a global economy.

The Pluriannual Plans implemented in the two presidential terms of Luiz Inácio 'Lula' da Silva represent a sequel to the previous era's major investment projects. The 2008–2011 PPA drafted the Growth Acceleration Program (PAC in Portuguese), which is a policy for major investments in infrastructure projects. Its main focus remains competitive global insertion through territorial transformation. The project was launched during the global financial crisis of 2008/2009. It proposed the planning and execution of large public works projects for social, urban, logistical and energy infrastructure. One of the objectives of PAC has been to guarantee the continuity of goods and services consumption; a major premise of the 2008–2011 plan was the continued expansion of the mass market (with policies to that end implemented by Cardoso). To do so, it was necessary to equip all regional territories with modern engineering systems, thus enabling them to absorb increased domestic and/or foreign flows of goods, people or information. The second phase of PAC, initiated in 2011 by Dilma Rousseff's government, emphasized 'biofuels'. One of the largest projects of PAC II, currently being executed, is the Ethanol Logistic System. This project will unite areas that produce sugarcane ethanol in municipalities in the states of Goiás, Minas Gerais and São Paulo. It will thus enable pipe transportation of the ethanol produced in rural areas to distribution centres and markets inside and outside the country.

These territorial planning actions confront the current global 'imperative of fluidity' (Santos 2002). To Santos, a major characteristic of the contemporary world is the demand for fluidity in the circulation of information, goods and finance for the ultimate benefit of international productive and financial capital. Fluidity, in turn, is based on what the author calls 'technical networks', which he presents as the material foundations of competitiveness: highways, ports, airports, gas and oil pipelines, canals, and networks for digital data transmission. These networks also add value to the activities that use them. Territorial planning in Brazil, through the country's various PPAs, has focused on constructing such technical networks across diverse territories and regions.

Santos (2002) also observed that producing fluidity is usually a cooperative process between public and private sectors. The state, directly or by means of concessions, has a duty to equip different territories with technical macro-systems; private companies, together or separately, establish networks with spatial features corresponding to their own market interests. The Ethanol Logistic System is an example of the cooperative process: the project had direct funding from the Brazilian Development Bank (BNDES in Portuguese)—using public resources— and the project execution, although officially a duty of the state-owned Petrobras, has been shared among the following private companies: Copersucar, Raízen, Odebrecht Transport Participações, Uniduto Logística and Camargo Corrêa Construções e Participações. Additionally, the direct users of the network are not only traditional sugar/bioenergy producers, but also enterprises associated with the fossil fuel sector, such Shell, as well as trading companies and agricultural businesses that have recently joined the biofuel market, such as the Bunge Group, Louis Dreyfus Commodities and Archer Daniel Midlands.

7.4 A Non-Conclusion

In this essay I have discussed some of Santos's theories relating to territorial planning and presented two distinct phases of territorial planning in Brazil from its integrationist and developmental origins to the contemporary period of global competition, networks, production, trade and information circuits. The long, ongoing history of territorial planning in Brazil is difficult to summarize. It is indeed possible, however, to identify links between the territorial planning of yesterday and of today. For instance, both phases focused on facilitating large-scale regional economic projects. I highlight two other linkages below:

- In both moments, the federal government assumed the duty of building technical macro-systems (that is, networks for circulation, transportation, energy and communications) and of making direct investments in production (that is, facilitating the presence of industrial companies in key strategic areas, such as the steel sector). In the first period, the state acted as a monopoly, connecting the national territory with the market and creating conditions for the diversification of industrial production. In the present period, large investment projects remain on the agenda of territorial planning policies, but their implementation is now shared with the private sector; large infrastructure is most frequently used by transnational firms—established in the country since the economic opening of 1990—rather than by the general Brazilian population. Some networks are, in fact, used exclusively by private corporations. In other words, large state-funded physical infrastructural projects were historically used to integrate regional subnational economies so that they could form a national market; later, these projects were in service of sub-national regional economies strongly tied to national circuits and, increasingly, international circuits of capital.
- Territorial planning in both periods has used rhetoric of reducing regional inequalities and poverty. However, both periods have also used planning policies that are based on spatial selectivity. In the past, income differences within the population and between regions constituted limiting factors to the expansion of capitalist modernization in the territories of peripheral economies. Today, in spite of the focus on the 'inclusion by consumption' as a measure of poverty mitigation, current territorial planning policy advocates for the inclusion of places and regions into the global market through the logic of selectivity. To this end, policy promotes different instruments of economic development for each city or region (or even city-regions), creating the conditions for interterritorial competition for public and private investments in order to attract firms, consumers and labour forces. In this context it may be asked: is planning fulfilling its original function of adding value, or is it exacerbating spatial differences through capital investment and profit, causing some territories to reach more economic prosperity than others?

Given these considerations, a question arises: is an alternative form and goal of territorial planning possible? As Santos (2003) argues, one need not qualify planning as *capitalist* because it has never been anything but. However, before promoting a new type of planning, which may be, variously, 'critic', 'participative' or 'democratic', we must deconstruct and understand, through careful analysis, the types of planning being implemented now: planning that has been, from its inception to today, articulated with capitalism. Santos helps us do just that.

References

Araújo, T. B. (2000). *Ensaios sobre o desenvolvimento brasileiro: Heranças e urgências.* Rio de Janeiro: Revan.

Cima, E., & Amorim, L. (2007). Desenvolvimento regional e organização do espaço: Uma análise do desenvolvimento local e regional através do processo de difusão de inovação. *Revista FAE, 10*(2), 73–87.

Egler, C. (1995). Questão regional e gestão do território no brasil. In I. E. Castro, P. C. Gomes, & R. L. Corrêa (Eds.), *Geografia: conceitos e temas* (pp. 207–238). Rio de Janeiro: Bertrand Brasil.

Lencioni, S. (2006). Reestruturação urbano-industrial no Estado de São-Paulo: A região da metrópole desconcentrada. In M. Santos, M. A. A. de Souza & M. L. Silveira (Eds.), *Território: Globalização e fragmentação* (pp. 198–210). São Paulo: Hucitec.

Lessa, A., Couto, L., & Farias, R. (2009). Política externa planejada: Os Planos Plurianuais e ação internacional do Brasil, de Cardoso a Lula (1995–2008). *Revista Brasileira de Política Internacional, 52*(1), 89–109.

Lima, J. F. (2003) A Concepção do espaço econômico polarizado. *Interações: Revista Internacional de Desenvolvimento Local, 4*(7), 7–14.

Piquet, R. (2007). *Indústria e ordenamento territorial no Brasil contemporâneo.* Rio de Janeiro: Garamond.

Santana, M. (2011). Planejamento territorial e subdesenvolvimento: Fatos e mitos. *Revista Geográfica de América Central,* Número especial XIII EGAL, 1–24.

Santos, M. (2004 [1974]). *O Espaço dividido. Os dois circuitos da economia urbana nos países subdesenvolvidos.* São Paulo: Edusp.

Santos, M. (2003 [1979]). *Economia espacial: Críticas e alternativas.* São Paulo: Edusp.

Santos, M. (2002 [1996]). *A Natureza do espaço: Técnica e tempo; Razão e Emoção.* São Paulo: Edusp.

Chapter 8
Rethinking Federalism through the Work of Milton Santos

Fabrício Gallo

Abstract The geographical analysis of a national territory requires a research method that considers the multi-faceted nature of reality, especially for those territories where state organization occurs within a federal structure. This structure uses a policy architecture to resolve the problem of the organization of power in the face of specific regional diversities of each national territory. The concept adopted in this chapter is of the *used territory*, as proposed by Milton Santos. The hegemonic agent par excellence in the use of territory is the state. In countries of federative state organization (in other words, where the state's power is shared between federated entities of isonomic form) the national public budgets are often characterized by the transfer of tax funds from one federated entity to another. The geographical analysis of these transfers is one way to understand how the state uses the territory. To exercise power, the state uses the territory by imposing norms that regulate and create tensions among entities because some transfers can favour only one group of subnational entities. This brings into question the principle of federal isonomy and can lead to a *war of places*, animated by disputes for limited public resources. This chapter discusses how Milton Santos's dialectic between *normed territory* and *territory as norm* can be used to interpret the taxation disarrangement in the federation.

8.1 Introduction

The geographical analysis of a national territory requires a research method that considers the particularities inherent to each territory. This text seeks to analyze territories in which the state is organized in a federal structure. In *Political Geography*, Sanguin (1977) has argued that a federal state implies an alliance, contract or pact among diverse regions and populations. The federal government

Fabrício Gallo, Assistant Professor, Department of Territorial Planning and Geoprocessing, Universidade Estadual Paulista—UNESP, Brazil; Email: fgallo@rc.unesp.br.

© Springer International Publishing AG 2017   91
L. Melgaço and C. Prouse (eds.), *Milton Santos: A Pioneer in Critical Geography from the Global South*, Pioneers in Arts, Humanities, Science, Engineering, Practice 11, DOI 10.1007/978-3-319-53826-6_8

regime is a phenomenon of constitutional law whereby the authority of the state is not hierarchized, but shared between a central government and other subnational levels. Each level, within the limits of its competence as defined by the constitutional text, performs the plenitude of the state's power. As such, federalism is a political architecture that aims to solve the problem of the organization of power in the face of specific regional diversities in each national territory. Its particular challenge consists of finding solutions for governmental questions in a complex relation of spatial differences and similarities.

In this essay I use the ideas of Milton Santos to analyze the geographical expression of federalism by interpreting it as an *event* in the process of uses of the territory (Santos 2002). In analyzing the Brazilian federation, I consider the relevancy of the concept of *used territory* (Santos 1999, 2002; Santos et al. 2000; Silveira 2009) to include all of the agents that constitute this federation in a totality that is divided into two matrices constituted by political actions: on the one hand, materialities, and on the other hand, immaterialities.

In this essay I interrogate how the concept *used territory* is important for the analysis of the federation; it helps me consider the power of each agent to use a given territory in accordance with his or her specific set of interests. I also propose understanding federalism as a geographical event because to do so presents federalism as having roots in an earlier moment in history, even though it exists in the current period. Lastly, I suggest that the unequal process of resource distribution in a federation can result in a selective renewal of materialities in the territory, resulting in a deepening of socio-spatial inequalities.

8.2 The Relevancy of the Used Territory in the Analysis of the Federation

The category *used territory* has two interrelated dimensions. First, it refers to *territorial configuration* (Santos 1988), from the smallest technical and natural objects to the largest engineering works. Second, it includes all social actions, from the repetitive actions that constitute daily life to actions that have the power to change the course of history. To Santos, then, *used territory* is an analytic category that places political actions and materialities side by side as co-constitutive. The actions advance the forms—the materialities—and the latter, in turn, condition social actions. In other words, space is animated by social forces and thus might be considered a social conditioner (Santos 2005). Santos et al. (2000: 104) continue by arguing that, 'used territory is constituted as a whole complex that consists of a web of complementary and conflicting relations'. Silveira (2009: 129) elaborates that: 'Used territory captures past actions, already crystalized in objects and rules, and present actions, which happen in front of our eyes'.

To Santos (2005), *territory* is composed only of forms, but *used territory* includes objects and actions: it is a synonym for human space or inhabited space. That is why Santos et al. (2000) consider *geographical space* not as synonymous with territory, but with used territory. The interpretation of a given geographical space requires distinguishing between the agents involved in action (in other words, those who have the power to use the territory), because used territory is constituted through the interdependence and inseparability of materiality and its use.

Hegemonic agents impose their use strategies on space, while non-hegemonic agents use the territory in other ways, with other intentionalities. The state is a hegemonic agent that wields immense power to define a territory's use. The state is the agent that commands the most coercive means, one of which is the power to create laws. Santos (2002) argues that the state, because of its prerogative to legislate, is the primary force capable of causing events to occur over extensive areas. Santos asks:

> What forces are able to cause events to occur, in a moment, over extensive areas? The first of them is the state, by its "legitimate use of force", within the law or not. Law...is, naturally, general. Thus, a public rule acts over the totality of the people, companies, institutions and territory. This is the superiority of state actions over other macroorganizations. (Santos 2002: 152)

The state is a force that produces events throughout a territory. Analyzing territorial transformations thus always demands understanding the political sphere in which decisions are made regarding the use of the national territory.

8.3 Federalism as Geographical Event

Using the concept of the *event* can help elucidate 'social actions', the latter which are co-constituted with a territory's materialities. Through events, historical and natural time are incorporated into geographical space. Thus it is possible to heuristically identify natural events (such as rain or earthquakes), which result from nature's dynamics, and historical events (such as classes, seminars, revolutions or coups d'état), which result from social relations, that is, human actions that cause people, with all of their contradictions, to interact. Both natural and social events, however, are the result of both human and territorial action.

Events result from a set of vectors (such as new social rules and laws, new political conditions or new technologies) that lead to new functions in the preexisting milieu: they reorganize some functions and disorganize many others. However, an event only takes place when it integrates, that is, when it materializes in geographic space.

According to Santos (2002), the duration of an event refers not only to a time fragment or the time course of an action, but also to its spatial extension. The duration and extension of an event depends on an organizational system that authorizes its permanence. This is the case for the geographical space of federalism:

its territory is the result of all state actions, which define, downward, the political behaviour of the subnational scales. This direction of influence is demarcated in covenants within the national constitution. The existence of these constitutional rules and pacts, which are continually remade, is necessary for federalism to continue as a system of state organization that organizes events in its territory.

Drawing from Santos's conceptualization of an event, it is possible to interpret a federation as an event at the national level because its scale of realization is the totality of the national territory (Gallo 2014). Viewing a federation as an event requires interpreting each national territory as a distinct *sociospatial formation* (Santos 1977).[1] In each, the spatialization of an event is shaped by the preexisting territorial realities. This interpretation is inspired by Santos (1977): to him, the concept of sociospatial formation attempts to undo a longstanding notion of society and space as dissociated. In seeking to understand economic formation processes (which are the result of social relations) and the spatial formation of a national territory, it is necessary to regard them as inseparable. As economic and social formations come into being, they are concretized in space, but this concretization only occurs through its adaptation to the rules of a nation-state. Therefore, *sociospatial formations* are manifest through the national territory—the territory of the nation-state—and can be defined as the territorial-state scale. Because economic and social formations are necessarily adapted to, and transformed by, the specific rules of each national territory, each sociospatial formation is distinct from those in other national territories. Thus, each national federation is also made distinct. Moreover, the sociospatial formation of a national territory shifts as federations are remade in each historical moment, following the particular 'block of power' (Poulantzas 1973) that commands the territory, which then uses this territory according to the block's determined order. This order is negotiated: the relation among the different entities within a block is constantly shifting, simultaneously transforming the use of a territory.

Brazil, like all nation-states, has a specific sociospatial formation. Moraes (2002) argues that after Brazil's political emancipation on 7 September 1822, Brazilian elites began to establish the new state in a vast, resource-rich territory that had not yet been consolidated through a hegemonic economic form. In this territorial formation, the country was not conceived as a *people*, but as a *land*, that is, not as a community of individuals but as a spatial territory. The conception of federalism in Brazil matured in accordance with the interests of elites to preserve their space of

---

[1]Regarding the notion of socio-spatial formation, Santos (1977) explains that for a long time the concepts of society and space were separated. Santos argues that economic formations (which are the result of social relations) and spatial formations of a country must be understood as inseparable from a theoretical point of view, and he created the category *socio-spatial formation* to capture this interdependency. That is, the economic and social formation materializes in space. But because socio-spatial formation is related to the mode of production, its realization is only possible through adaptation to the laws of a nation-state. Thus the category socio-spatial formation can be defined as the scale of the nation-state, ie the national territory. These attributes emphasize the importance of the specificities of each territory.

domination, in the midst of tensions derived from a strong system of social exclusion and the threat of disorder that arose from the weakening of the slave regime. The political conditions for a federative organization of the Brazilian state arose specifically after the Proclamation of the Republic (on 15 November 1889) by the Promulgation of the Republic of the United States of Brazil's Constitution (on 24 February 1891).

From a territorial point of view, federalism in Brazil was a political mechanism designed to solve two problems: (i) regionalism, which during the monarchy (from 1822 to 1889) threatened to fragment the territory; and (ii) the lack of physical land networks which could integrate the extensive territory making communication between the parts of the Empire more effective. This set of features, which were in existence before the establishment of federalism, is *unique* because it is specific to the Brazilian territory. Thus it is useful to understand Brazilian federation as an *event,* occurring in time, that reorganizes the territorial formation. As Santos explains (2002: 163): 'Events are individual, but there are no isolated events. They are interrelated and interdependent, and it is these conditions that participate in situations.'

The establishment of federalism in Brazil represents a new territorial formation. According to Santos (2002), a *ruled territory* is a national territory ruled by laws that make it formally distinct from other territories. This understanding of territory is consistent with Gottmann (1975), who argues that territory is a political compartment surrounded by frontiers that is validated by a juridical body distinct from the territories that surround it. However, places differ in their respective characteristics, that is, by the set of singularities that make local realities diverse and unequal. Santos (2002) described this as *territory as rule*, or territory as a norm: here, territory is an active participant in the way that certain actions are manifest in a place. In the Brazilian case, federalism as an event was a territory imposing itself as a rule because the territory was a 'barrier' to the performance of centralized power —it inspired federalism.

An event is formed by multiple forces in permanent conflict. The imposition of an event, such as federalism, results from combinations of contradictory social forces. These contradictory forces take shape as a dialectic between *rule* and *ruled,* which is a manifestation of a territory's transformations that require a rearrangement of the social and political organization of the nation.

8.4 Selective Renovation of Materialities and the Distribution of Tributary Resources Among Federate Entities

As discussed above, the purpose of this text is to understand the federal structure as the broadest analytic segment in the *used territory* of Brazil. In order to understand the politics of the used territory it is necessary to attend to the constant negotiations among different entities for intergovernmental transferences of tributary resources.

In Brazilian federalism, transference mechanisms are of great importance because they allow greater or lesser local and regional autonomy; some intergovernmental transfer mechanisms allow portions of the territory to receive more resources than others, thus providing a *selective renovation of materialities*. In other words, only some parts of the territory are selected by corporate interests to be modernized, while most of the territory remains in a precarious condition (Santos & Silveira 2001). Once resources are transferred to more privileged regions, they become embodied in specific places as materialities, increasing the *technical density* (Santos 2002) of portions of the territory. A territorial interpretation of the Brazilian tax system reveals that federalism is central to the selective spatialization of transferred resources and the resources' resultant materialities (because new infrastructure— such as streets, roads, communication networks and electricity, urban sanitation, schools and health centres—are implanted in the territory). This type of analysis avoids economism. Economistic analyses lack a territorial perspective as they only consider the equitable distribution of resources deduced from formulations and mathematical indices. A territorial perspective offers the ability to reduce sociospatial and infrastructural inequalities.

The Brazilian Federal Constitution of 1988 guaranteed isonomy among entities and the autonomy and exclusive competence to make certain decisions that are constitutionally established as proper for each entity (these include, for example, establishing that: the Union is responsible and has the autonomy to organize the armed forces, immigration policies and macroeconomic policy; the states are responsible for metropolitan regional policies; and municipalities are responsible for the organization of municipal routine, and the legislation and organization of cities). However, these exclusive competencies are also extended to tributary matters and thus each entity has the exclusive competence to tax certain fields. For example: the federal Union collects income tax from each individual or company, tax on industrial production, fuel tax, tax on import and export, financial transaction tax etc.; the states are responsible for the collection of taxes on production and circulation of merchandise and transport and communication services, vehicle property tax etc.; and municipalities are responsible for collecting taxes on urban territorial property, on services performed etc. Because it has control over the entire national territory, the federal government is the entity that collects the most taxes. There are intergovernmental mechanisms that attempt to equalize this resource distribution, such as: (i) constitutional and legal transferences (with rules established in the Federal Constitution and Federal Laws and Decrees) pursuant to which all entities are assured their portions of resources from other entities; and (ii) voluntary transfers, which include agreements and several actions that have been formalized between the federal government and states and cities, as well as between states and cities (Brazil-CGU 2005). For the 2016 Olympics Games, for example, the Brazilian federal government signed a contract with the municipality of Rio de Janeiro to release approximately US\$ 148,000,000 for the construction of the VLT—light rail vehicle—in the city. In this latter transference modality—the voluntary transfers—all entities can demand resources. However, this does not guarantee the formalization of the agreement or the release of money pursuant to an agreement that was previously signed, because there is not money for everyone. Agreements cannot always be fulfilled because of shifts in budget frameworks (for example, a tax revenue reduction caused by a crisis); backstage political negotiations thus often occur amidst budget deals.

This process exemplifies work by Affonso (1995), who argues that the Brazilian federation stems from a complex set of alliances, usually not explicit, forged largely by public funds based on fiscal policies that were adopted throughout the years. A synthesis of the Brazilian budget process is described by Rezende and Cunha (2002), who affirm that this process can be seen as a game with many participants including the president, ministers, deputies, senators and hundreds of members of the technical teams of the three powers (Executive, Legislative and Judiciary), plus interest groups articulated with companies, associations, federations, confederations and state and municipal governments. All try to include, maintain or expand their share of benefits provided by budget funds. In the conception of Arrais (2008) the resource transfers are important elements in the national political arena, as they are the product of arrangements that frequently go beyond legality. Thus, it is important to map and understand this distribution of resources as a component of power relations.

These transfers establish new infrastructure or restore existing infrastructure, selectively strengthening the *technical density* of various federate municipal entities in the Brazilian territory. Santos's (2002: 306) concept of used territory draws attention to how the emergent material or infrastructural profile is 'the base of the use values and exchange values…One can say, considered in their technical reality and in their use rules, infrastructures 'rule' behaviors and then 'choose' or 'select' the possible actors'. This is how *territory appears as rule*: it participates in the selection of certain actions and actors that will appear at the local scale. Put another way, many of the new materialities to be installed by transfers require pre-existing infrastructure that is only found in some parts of the country. As a result, those territories with pre-existing infrastructure have agency in selecting actions and actors—this is the territory imposing itself as rule.

The diffusion of modernization is responsible for the aforementioned *selective renovation of materialities* (Santos & Silveira 2001) in the national territory. Large engineering works and modern technical objects and facilities for the achievement of citizenship (such as hospitals, schools, day care centres, universities, water and sewage treatment) tend to be distributed to places. However, distribution choices also stem from the structure of the federation, in other words, the pacts and political negotiations between various entities, which are not always clear to the rest of the nation. This means that even if the federal government calls for transference proposals, and even if the municipalities submit projects with all of the correct documentation, there is no certainty that an agreement will be made, because there is not enough money for everything and because of the often obscure negotiations between public and private agents that influence the granting of money (Gallo 2013).

8.5 Final Considerations

In this text I have explained the political engineering of Brazilian federalism through one of its territorial expressions: resource transference. The chapter has focused on how the distribution of resources among entities occurs and is materialized within the territory. Following Santos, a full understanding of all norms, rules, laws, conditions and articulations, alongside a full understanding of how they have been produced by *concrete politics* and how they have produced a *concrete federation,* is one way for individuals in the municipalities to discuss and contribute to the decisional processes for the future of their territories. A full understanding of how real politics and real federations are produced is important considering that many of the negotiations between the agents are obscure to most Brazilian citizens. Due to this obscurity, there is a use of the territory and a process of renewal of materiality that caters predominantly to hegemonic interests and not the entire national society.

I argue that this federation can be interpreted both as a formal federation and as a concrete federation. The former, the formal federation, is one that is established by legal and juridical structures where all entities should have the same rights and potentialities for development within the nation. A federation in fact is one that is constituted by agreements and political alliances that have a territorial basis, where movement turns into political order and where *territory imposes itself as rule,* that is, a federation where parts of the territory are developed more than others because of competing corporate interests and pre-existing infrastructure. The federate principle of isonomy between entities is at risk and deep socio-spatial inequalities become feasible: some portions of the national territory always appear to receive more funds for the establishment of new infrastructure and increases in their *technical density* because of the central role of local politicians in the negotiation of agreements. However, the process occurring at this specific policy conjuncture in Brazilian society—in which some portions of the territory indirectly impose themselves over others—has in fact been occurring since the origin of the Brazilian federation at the end of the 19th century.

Thus, following Santos, I suggest that the federation can be considered as a *geographical event* that continues in its *duration* up to the present. As Santos (2002) states, the event is always present, but its duration is a period in which the event has been repeating with its original characteristics. The Brazilian federation, which was established in 1891, has a duration that has extended up to the present time: it continues to be characterized by elements that were present at its origin (agreements, negotiations and pacts accomplished between political class and private agents), but it is also being shaped by and adapting to the political conditions of today.

References

Affonso, R. B. A. (1995). A federação no Brasil: Impasses e perspectivas. In R. B. A. Affonso & P. L. B. Silva (Eds.), *A federação em perspectiva: Ensaios selecionados*. São Paulo: FUNDAP.

Arrais, T. P. A. (2008). Diversidade territorial e transferências constitucionais para os municípios: Considerações sobre a economia regional goiana. *Boletim Goiano de Geografia, 28*(2), 203–216.

Brasil – Presidência da República, Casa Civil. (1891). *Constituição da República dos Estados Unidos do Brasil, de 24 de fevereiro de 1891*, series. Retrieved August 30, 2016, from http://www.planalto.gov.br/ccivil_03/constituicao/Constituicao91.htm.

Brasil – Presidência da República, Casa Civil. (1988). *Constituição da República Federativa do Brasil. Texto constitucional de 1988*, series. Retrieved August 30, 2016, from http://www.planalto.gov.br/ccivil_03/constituicao/Constituicao.htm.

Brasil – Controladoria-Geral da Federal Government (CGU). (2005). *Gestão de recursos federais. Manual para os agentes municipais*. Brasília-DF, Secretaria de Controle Interno, series. Retrieved August 30, 2016, from http://www.cgu.gov.br/publicacoes/auditoria-e-fiscalizacao/arquivos/cartilhagestaorecursosfederais.pdf.

Gallo, F. (2013). Território, política e infraestruturas: A influência do governo federal na política urbana dos municípios brasileiros. *Sociedade & Natureza, 25*, 453–467. Retrieved August 30, 2016, from http://www.seer.ufu.br/index.php/sociedadenatureza/article/view/2298.

Gallo, F. (2014). Elementos da Formação Territorial Brasileira. A Federação Nacional como Evento Geográfico. *Boletim Campineiro de Geografia, Campinas, 4*(2), 27–43. Retrieved August 30, 2016, from http://agbcampinas.com.br/bcg/index.php/boletim-campineiro/article/view/167/2014v4n1_FGallo.

Gottmann, J. (1975). The evolution of the concept of territory. *Social Science Information sur les Sciences Sociales, 1*(3/4), 29–47

Moraes, A. C. R. (2002). *Território e história no Brasil*. São Paulo: Hucitec/Annablume.

Poulantzas, N. (1973). *Political power and social classes*. London: New Left Books.

Rezende, F., & Cunha, A. (Eds.). (2002). *Contribuintes e cidadãos. Compreendendo o orçamento federal*. Programa de estudos fiscais. Rio de Janeiro: FGV/IBRE: FGV/EBAPE.

Sanguin, A.-L. (1977). *La géographie politique*. Paris: Presses Universitaires de France.

Santos, M. (1977). Society and space: Social formation as theory and method. *Antipode, 1*(9), 3–13.

Santos, M. (1988). *Metamorfoses do espaço habitado: Fundamentos teórico e metodológico da geografia*. São Paulo: Hucitec.

Santos, M. (1999). O Território e o saber local: Algumas categorias de análise. *Cadernos IPPUR, 13*(2), 15–25.

Santos, M. (2002 [1996]). *A Natureza do espaço: Técnica e tempo; Razão e emoção*. São Paulo: Edusp.

Santos, M. (2005). O Retorno do território. In M. Santos (Ed.), *Da totalidade ao lugar*. São Paulo: Edusp.

Santos, M., & Silveira, M. L. (2001). *O Brasil. Território e sociedade no início do século XXI*. São Paulo: Record.

Santos, M., et al. (2000). O papel ativo da geografia: Um manifesto. *Território, 9*, 103–109. Retrieved August 30, 2016, from http://www.revistaterritorio.com.br/pdf/09_7_santos.pdf.

Silveira, M. L. (2009). Ao território usado a palavra: Pensando princípios de solidariedade socioespacial. In A. L. A. Viana, N. Ibanez & P. E. M. Elias (Eds.), *Saúde, desenvolvimento e território*. São Paulo: Aderaldo & Rotschild.

Weber, M. (1978). *Economy and society: An outline of interpretive sociology*. Berkeley and Los Angeles: University of California Press.

Chapter 9
Milton Santos's Contribution to Understanding the Transformations Underway at Modern Agricultural Frontiers

Júlia Adão Bernardes

Abstract The work of Milton Santos offers a wide range of concepts that may be used to analyze modern agricultural development in Brazil. This chapter discusses how these concepts, particularly that of space as a *system of objects* and *system of actions*, can be useful for analyzing transformations of modern agricultural frontiers in the Brazilian states of Mato Grosso and Pará. Such transformations are intrinsically linked to the intensification of activities along the BR-163 federal highway, which connects the southern and northern regions of the country. The text analyzes the sociospatial impacts of this nation-state infrastructure and of the reorganization of agricultural territory along these new frontiers.

9.1 Introduction

Modern agribusiness in Brazil, connected to the ebbs and flows of globalization, is creating new geographies of production along the BR-163 federal highway[1] through Mato Grosso and Pará. As Milton Santos explains:

Júlia Adão Bernardes, Professor, Department of Geography, Federal University of Rio de Janeiro, Brazil; Email: julia.rlk@gmail.com.

[1]The construction of the BR-163 was initiated in 1973 during the Brazilian military dictatorship—Medici's mandate—in the context of the First National Development Plan (1st PND, 1972–1974). In its first phase the highway connected the cities of Cuiabá in the state of Mato Grosso and Santarém in Pará. BR-163 construction took place during the same period as the construction of the Trans-Amazon Highway. These new works were also accompanied by a military integrational plan for the construction of roads with the twofold goal of: (i) geopolitical control of the Amazon; and (ii) stimulation of economic and agricultural development.

© Springer International Publishing AG 2017   101
L. Melgaço and C. Prouse (eds.), *Milton Santos: A Pioneer in Critical Geography from the Global South*, Pioneers in Arts, Humanities, Science, Engineering, Practice 11, DOI 10.1007/978-3-319-53826-6_9

[These] are the new fronts that are born technified, scientified, informationalized...If the pioneering movement of São Paulo...was led by the great cultivators capable of building railroads, attracting immigrants, and taking advantage of modern machinery, today the pioneering fronts are opened primarily by big business with the cooperation of the authorities. (Santos & Silveira 2001: 119)

The fundamental connection between technology, science and information, expressed through scientification and technologization of the landscape, is producing modern agricultural frontiers.

Milton Santos's work offers a broad array of concepts that can guide research on this spatialization of agricultural processes. His work also allows us to interpret the effects of these new technologies and rationalities for people who live in regions undergoing rapid market expansion. This chapter explores the relevance of Santos's concepts for analyzing the expansion of the modern agricultural frontier[2] in the Brazilian state of Mato Grosso, in the midwest region of the country, and Pará, in the northern region. This frontier is constructed through a system of objects and a system of actions, a technosphere and a psychosphere, and space and society. These concepts comprise what Santos calls the technical-scientific-informational milieu.

Santos's analytical approach requires recognizing the complexity and creation of new territorialities in the current phase of globalized capitalism. This phase is marked by: new technical and territorial divisions of labour, market dynamics and state actions; clashes between dominant interests and pre-existing production methods; and the juxtaposition of different times and spaces of social struggles. Santos's approach allows us to understand how the reproduction of social relations occurs through the adaptation of space; by drawing links between action, time and space, the agricultural frontier can be mapped and understood as, first, a project pursued by those who have the most economic and political power, and, second, as a spatiality where people struggle for their livelihoods within the prevailing hierarchical, excluding order. At a moment of accelerated flows in the capitalist world, Milton Santos's theories provide important concepts for understanding processes of capitalist accumulation. The analyses in this text demonstrate how Santos's theoretical framework and conceptual tools can be applied in the present day.

This chapter begins by describing the new characteristics of the recent agricultural frontier in Mato Grosso. Next, it discusses how Santos's concepts, particularly that of space, can be useful for analyzing transformations of modern agricultural frontiers. It then uses these concepts to interpret the particular agricultural frontier in Mato Grosso. The text concludes with a discussion of the sociospatial impacts of the new frontiers in both the state of Mato Grosso and in its neighbour, the state of Pará.

---

[2]In the Brazilian geographic literature the term agricultural frontier (*fronteira agrícola*) refers to areas of expansion of the production of new agricultural or livestocks products. Originally it referred to areas where native habitats were converted into agriculture. However, the term today also refers to areas where one type of production is replaced by a new one.

9.2 New Expansion of the Modern Agricultural Frontier in the Brazilian State of Mato Grosso

The first decade of the 21st century saw the emergence of new capital accumulation strategies in Brazilian agriculture under the country's new economic and financial policies (Delgado 2012). Statistics on the growth of its main commodities are illustrative of the policies' effects. From 2000 to 2012 the country's soybean production rose by 101%, while maize production increased by 120%, cotton rose by 148%, and the number of broiler chickens grew approximately 57%. The area of land given over to soybean farming increased by 83%, while 19% more land was planted with maize and 75% more with cotton.

Mato Grosso is often referred to as the 'world's bread basket'. The state has large tracts of relatively cheap acreage and suitable conditions for growing grains and producing meat, with large areas of sparsely populated flat land. The BR-163 federal highway running through Mato Grosso has accelerated the expansion of the agricultural frontier. Certain agribusiness sectors, like soy, maize and meat, have experienced substantial growth due, in part, to an intense consolidation of capital through mergers and joint ventures and recent transportation infrastructural development. The statistics demonstrate this rapid expansion: in 2012 this state was Brazil's number one soy producer, accounting for 31.17% of Brazilian output. From 2000 to 2012 Mato Grosso's production of soy, maize and cotton rose by 149%, 994% and 180%, respectively, while the total number of broiler chickens rose by approximately 226%. These figures are the combined result of boosted productivity and physical expansion, with the area of land occupied by soybean, maize and cotton plantations rising by approximately 140%, 387% and 180%, respectively, in the same period.[3] Agriculture has thus become a competitive advantage of Mato Grosso.

The federal state took a series of actions in 2000 designed to attract new agents to the region and thus accumulate more capital. These actions included creating new logistical systems for transporting produce northward[4] and introducing new development hubs that have expanded the region's so-called 'export corridors'. The federal government has also cultivated agribusiness development through credit policies and tax breaks. These actions have resulted in shifts in land use and land structure. The countryside has become extremely vulnerable to large capital ventures as its land has become a new resource for big business. Capital, land and labour are key elements in this expansionary movement because they are the 'organizing principle of society' (Polanyi 2000: 97). They unite to shape Brazilian agricultural production: money in the form of grandiose investments, land for large-scale projects and labour as the most valuable resource for extraction are all found in Mato Grosso.

The contemporary field of agribusiness is organized around the interests of new hegemonic actors and global forces, with science, technology and information

---

[3]Data from the Brazilian Institute of Geography and Statistics (Instituto Brasileiro de Geografia e Estatística, IBGE).

[4]Such as the creation of the hydroway Tapajós-Amazonas and the port of Mirituba (Pará).

underpinning the creation of new, efficient processes. This has resulted in a science-, technology- and information-intensive landscape, constituting what Santos calls the technical-scientific-informational milieu (Santos 1996). Power operates through this landscape as large-scale *action*. Action here is the use of time, which is fundamental to production processes. For instance, setting the crop calendar introduces greater temporal efficiencies (Bernardes 2007). Hegemonic actors are central in re-shaping time through introducing such actions.

New actions are thus shifting temporalities in this agricultural region of Brazil: technology is a precondition for efficiency, triggering processes that have created territorialities with new content, functions and structures.[5] For instance, technologies used in modern agriculture can improve average yields through the use of varieties that are genetically suited to the *cerrado* biome,[6] through fertilizers and pesticides, and through modern tools and machinery that assist in scaling up production (Bernardes 2011). The aim of my approach, drawing on Santos, is to consider action as having meaning within the ambit of capitalist rationality, linking it to time and space. In the next section I explore some of the Brazilian geographer's relevant concepts in greater depth.

9.3   Milton Santos on the Agricultural Frontier

To better understand the recent movement of the agricultural frontier in Mato Grosso, I deploy the categories of space and time as defined by Santos (1996). He argues that in the current technical-scientific-informational milieu, space is 'formed of an inseparable set of systems of objects and systems of actions' (Santos 1996: 51). A key aspect of a Santos-inspired analysis is to draw associations between action, time and space.

Actions must be synchronized, and their orders must comply with predefined times and interrelate with mechanisms that reproduce hegemonic power (Ribeiro 2002). Technology is central to this process. Santos (1996: 45) encourages us to perceive that: '[T]echnologies are involved in producing a perception of space, and also a perception of time, [producing] both their physical existence that marks the senses through speed, and their imaginary aspects'.

---

[5]Santos conceptualizes form, function, structure and process as four central 'internal categories' for understanding totality: 'There is neither structure nor function without form. Every form has a function, which serves both to cooperate with, and contradict, the structure. It is a question here of a form with a content, a form-content, an actuality, in opposition to the empty form, which is either an expectation or a delusion. The essential point is that the categories structure, function, form and process are inseparable both as analytical categories and as historical categories. They are the ones which define the concrete totality, the totality in its permanent process of totalization' (Santos 1980: 45).

[6]*Cerrado* is a vast tropical savanna biome present in the midwest region of the country.

Technologies re-order space and time. Modern technologies seek to master space and time, and ultimately it is the actions of capital that determine how technologies are used, creating new types of space and hence new relationships with time.

Santos considers space as a set of fixed elements and flows. Elements that are fixed in some places modify that place, as well as cultivate new or renewed flows that affect environmental and social conditions (Santos 1996). Santos distinguishes between different spaces as they are constituted through technology. He calls luminous spaces 'those that accumulate most technological and informational density, and are therefore better able to attract activities with greater capital, technology, and organizational content. Conversely, the sub-spaces where these features are absent are opaque spaces' (Santos & Silveira 2001: 264). There are many opaque spaces across Brazil that have little or no infrastructure, and are constituted by poor people and migrants (Santos & Silveira 2001).

Santos (1994: 83) understands time as created through humans: 'Time is given by humanity.[7] The concrete time of humanity is practical temporalization, the movement of the World inside each one, and thus a personal interpretation of Time by each group, each social class, each individual'.

But time, 'given by humanity', is also intimately related to technology: time in general is accelerated as a function of modern technology. But other, less modern forms of technology also exist that operate in slower temporalities. In Santos's conceptualization, the rich live in luminous spaces with plenty of present time and think only of stock prices, whereas the poor, in opaque spaces, are obliged to eke out a living in slow time. Santos describes slow people as those who do not master modern knowledge and thus can make a different territory that may bring about change.

Brazil is predominantly a country of the poor, the common person, slow people, migrants and opaque spaces, albeit with luminous spaces that punctuate its horizon. The slow people or the common people, however, are rapidly discovering the globalized world, with new life strategies and in solidarity with one another. Yet, through global market expansion, new technologies are speeding up time and creating different forms of space within this periphery. Slow and fast times, forming opaque and luminous spaces, are consequently being juxtaposed.

The action of production is organized horizontally in space. Natural conditions of a given space structure this action. For instance, in the case of soybeans, temperature, rainfall distribution, daylight and topography all structure agricultural output. However, socio-spatial factors also shape actions of production. For instance, the old forces and traditional values of past history have to be overcome for modern production and expansion of agriculture to happen. Santos (1996: 113) calls these forces *rugosity*: 'whatever is left over from the process of suppression, accumulation, superimposition.'

---

[7]Santos uses the term '*homem*' ('man' in Portuguese) as a synonym for humanity. Because he implied the term humanity—not limited to men—I have chosen to translate 'man' into gender-neutral form.

Importantly, spaces are linked to the global economic system. As Santos argues:

[I]n this phase of history, the world is marked by new signs, such as: the multinationalization of firms and the internationalization of production and products; the generalization
of the phenomenon of credit, which reinforces the features of the economization of social
life; the new roles of the state in a globalized society and economy; the frenzy of a
circulation that has become essential for accumulation; the great information revolution that
links places instantaneously thanks to the progress of information technology. (Santos
1994: 123)

Because the global economic system is competitive and globalized, it cultivates
highly technology-intensive corporate spaces. As such, space is fragmented so that
capital can penetrate different parts of the territory—those best suited for its
reproduction.

9.4   New Space-Times of Mato Grosso

Recent transformations in modern agriculture along the BR-163 highway in Mato
Grosso, leveraged by new logistics options, have created profound changes in the
roles and potentialities of the different spaces that constitute the region. Science,
technology and information have been essential to scaling up production.
According to IBGE data, from 2005 to 2012 the production of soybeans and maize
grew 55% and 334%, respectively, in the BR-163 integrated area of Mato Grosso—
the combined effect of a larger physical area and improved average yields due to the
technological developments associated with precision agriculture.

The new poultry, pig and cattle farming enterprises in the region are some of the
country's biggest grain and meat producers. Technology is central to how these
producers run their businesses: highly efficient farming practices, for example, have
contributed to the scale at which broiler chickens are being produced and slaughtered. In the year 2012, Sadia slaughtered approximately 300,000 chickens daily,
Perdigão approximately 375,000, and Anhambi 65,000. These companies also farm
high genetic quality pigs under strict sanitation standards. The number of heads of
cattle also increased approximately 30% from 2005 to 2012.

Land use in Mato Grosso is thus under the influence of new global market
dynamics, and luminous spaces are being formed in this region through selective
technology-based systems. The spatial design of production is changing to include
extensive areas devoted to the production of grains and meat with high added value.
The new globalized and informational dimensions of capital have prompted new
interactions between companies and places, establishing novel hierarchies between
different forms of capital and territories. Fixed elements[8] are being multiplied,

---

[8]According to Santos, geographic space can be considered an ensemble of fixed elements and
flows: 'Fixed elements, fixed in each place, allow actions that change places themselves, through
new or renewed flows that recreate environmental and social conditions and redefine each place.
Flows are a direct or indirect result of actions and cross or install themselves in the fixed elements,

diversified and renewed, and the flows are increasingly intense. Selected places are valued and specialized and circulation is accelerated. Santos explains this new spatiality:

> [W]ith the development of production forces, regional inequality ceases to be the result of natural aptitudes and simultaneously becomes more profound and more speculative: there is a growing need for increasing volumes of capital; social resources also tend to be concentrated in certain places where the productivity of capital is on the rise. Everything is connected. (Santos 2003: 22)

The regional division of labour is thus also shifting in response to the growing reach of the market, always in an interconnected fashion.

The creation of globalized agriculture and livestock farming networks, in conjunction with the fluidity enabled by new transportation and communication systems, has allowed circuits of production and circles of cooperation to reach increasingly distant areas. According to Santos (1994: 128), 'circuits of production are defined by the circulation of products, i.e. of material. Circles of cooperation associate other flows that are not necessarily material to these material flows: of capital, information, messages, orders'. Santos and Silveira (2001: 144) hold that capital ultimately brings together 'what the direct production process separated into different companies and places, through the appearance of real circuits of cooperation'. They stress that spatial circuits of production and circles of cooperation differentiate the use of space, generating more intensive, extensive and selective flows. These circuits constitute coherent systems that articulate agricultural produce with industrial products and related services. They have, as a result, spurred the reorganization of the countryside and the city and established new city/countryside relationships.

As a consequence new agribusiness activities are transforming both rural and urban space. New agricultural facilities are being introduced in the countryside, such as large piggeries and poultry houses. These transformations in the agrarian space are associated with industrial developments of the urban, including crushers, animal food manufacturers, warehouses, slaughterhouses, refrigeration units and other industrial facilities. In these urban spaces, sophisticated communication and information systems are also being introduced, reducing the time and redefining the spatial organization of production circuits. These systems have enabled space to be organized in networks that straddle municipal borders and allow more fluid flows of capital (Santos 1996).

To understand the dynamics of new spatializations in areas influenced by the BR-163 in Mato Grosso, it is crucial to understand movement through these circuits. In this region, for instance, new spatial circuits of production of grains and meats are constituted through the flows of these goods along with capital and information. Spatial differences along this highway route have a hierarchical

---

(Footnote 8 continued)

changing their signification and value, at the same time that flows also modify themselves' (Santos 1996: 50).

structure. Differences result from a combination of diverse variables—different levels of technology, production relations, profit margins, class struggles—accompanied by varying levels of capital investment and infrastructure. Generally speaking, the region located along this axis has different levels of productivity and accumulation of capital, with each hierarchical level corresponding to a specific function in the social and territorial division of labour (Bernardes 2011).

9.5 New Contradictions in Mato Grosso and Pará: Competing Values and the Exploitation of Labour and Natural Resources

In this extremely dynamic capital frontier, there is a speeding up of movements and a process of socialization driven by globalization's production and circulation systems. The result has been brusque changes in material life, massive environmental consequences and a denial of the positive contributions of local cultures (or of 'slow people', to use Santos's vocabulary). In this section I explore these contradictions as they manifest in both Mato Grosso and in the neighbouring state of Pará.

There is a northward movement of the modern agricultural frontier in Brazil, driven by the intensification of activities along the BR-163 and encouraged by the creation of new routes for transporting produce. This has resulted in the accelerated introduction of new fixed elements and actions in a different spatial segment of the highway in the state of Pará. Expansion involves the rapid relocation of agribusiness activities such as cattle farming. According to IBGE data, there were 4,960,794 heads of cattle in the BR-163 region in the state of Pará in 2005. In 2012 this figure had risen to 6,447,751, representing a 29.97% increase. By pushing livestock rearing northwards to Pará, the state of Mato Grosso has been able to turn land over for soybean farming.

Agribusiness has significant environmental consequences. Logging, deforestation and the planting of pasture are necessary precursors to cattle farming. As for soybean cultivation, even though the conditions for this crop are not ideal in Pará, the practice has gradually been introduced, occupying land from the forest biome, to take advantage of the new logistics options. As a consequence, there has been extensive deterioration of the physical and biotic environment. Prodes/INPE[9] data indicate that deforestation in the BR-163 area of influence in Pará rose by 17.54% between 2005 and 2012. This destruction has had a significant impact on cultural and ethnic groups who rely on this land: there has been a loss of identity for some native groups who have long lived in harmony with the prevailing ecosystems in the region.

---

[9]Brazil's official deforestation estimates are made by the Amazonian Deforestation Monitoring Project (Projeto de Monitoramento do Desflorestamento na Amazônia Legal, Prodes) run by Instituto Nacional de Pesquisas Espaciais (INPE).

New and old come into conflict through market expansion along these frontiers of the globalized economy. Martins (1997) describes the frontier as a new economic rationality based on the formal and institutional constitution of new political mediations. With economic expansion, new production spaces are formed that offer new potential and have multiple driving and conditioning factors. The new frontier, linked to the expansion of markets, induces modernization and new conceptions of life (Martins 1997). There is also a simultaneous diminishment of traditional spaces as values of rationally-oriented production are being introduced. For instance, utility is valued above almost all else; people, knowledge, institutions, places and nature are transformed into means and technological utilities on a rationally designed plane. The ultimate goal is reproducing the capital accumulation process.

There are also myriad effects of agribusiness on local labour practices. As the economic frontier moves, access to the land is curbed as a result of its increased value. In other words, land valuation hampers access to land. This results in a greater concentration of land ownership and the exclusion of small- and medium-sized landholders. Until recently, the economy in the area of influence of the BR-163 in Pará was based on mixed family farming, but now disputes over land are abundant.

When smallholders are deprived of their means of subsistence and autonomy, a shifting pool of unskilled labour is formed—people who normally have no fixed employment and therefore have to travel to find work and accept substandard working conditions. According to the Comissão da Pastoral da Terra, slave labour is particularly prevalent in logging and sawmill[10] sectors in these regions. This situation is testament to the lack of an egalitarian ideology at the frontier. Market expansion creates a class of individuals stripped of their means of (re)production.

As a consequence of the movement of this agricultural frontier, indigenous groups, who have been barriers blocking the advance of the technology frontier, are finding themselves trapped in indigenous reserves that act somewhat like 'preservation islands'. Despite their name, these islands are not immune to the spatial transformations in neighbouring areas occasioned by the destruction of the ecosystem. This destruction includes the extinction of animal species, which erodes these groups' quality of life because of the resulting shortage of food.

Like Ribeiro (2009), I believe we are facing a new potential interpretation of the frontier. The frontier can be understood as a spatialization of the collision between the velocity required by global drivers and the remnants of past social struggles. Here, the organization of the territory implies the unharmonious coexistence of different rationalities. The imposition of new rationalities and a technological order transforms the class experience. It is a situation that demands a new territorial order—more land and new resources.

---

[10]Comissão Pastoral da Terra (CPT) reported 37 conflicts over lands around the BR-163 in 2013, affecting 2,755 households. In the same year, the CPT received ten reports of slave labour involving 98 workers engaged in logging, sawmills, livestock farming and soybean farming.

Crucially, these collisions may offer hope. As previously discussed, Santos considers the potentialities of slow people—the inhabitants of opaque spaces. His writings suggest different potential uses for today's technologies and the construction of a different type of globalization through particularities, occasioned by these slow people. In Santos's view:

> Today's world also authorizes a different perception of history through the contemplation of the empirical universality constituted with the emergence of new globalized technologies and the possibilities opened up for their use. The dialectic between this empirical universality and particularities will encourage the overcoming of the inverted praxis, thus far ruled by the dominant ideology, and the possibility of overcoming the reign of need, opening room for utopia and hope. (Santos 2000: 168)

9.6   Concluding Remarks

This chapter has explored Milton Santos's reflections to interpret the current expansion of the modern agricultural frontier along the BR-163. Santos's theories help me analyze the expansion of modern agriculture at this phase of technological development and lead to an interpretation of territory in which growing productivity is associated with social costs. Santos importantly urges us to value a territory beyond its potential for accumulating capital, to fight for the inclusion of every person.

It is fundamental to encounter the territory of resistance. To do so, one must interpret discourses other than the hegemonic, and different from the single logic that is imposed from without by global capital. These are discourses that propose different forms of living, different kinds of cooperation, different interpretations of the resources within space. We must endeavor to understand the mechanisms of the new solidarity that challenges the perversity inherent to the rapid speed of competitiveness, not least because, as Santos (2000) assures us, if the poor people who constitute the majority cannot consume the globalized West in its economic, financial and cultural form, they will end up relativizing or even rejecting globalization.

Santos (1996) indicates that it is necessary to penetrate the mystery of forms, ignored by many geographers. He argues that we must move beyond the false objectivity of the world of the senses. We must not interpret things as merely things, space as merely space, but instead seek their essence.

In order to interpret the territory of the modern agricultural frontier along the BR-163, to penetrate the mystery of its forms, it is necessary to perceive that technological, scientific and informational developments can potentially constitute a future whose actions diverge from the plans of the hegemonic players. It means developing alternate models from unique features, overcoming mere commercial relations to attain a higher state of cooperation. To develop these models, different segments of society must recognize that there are other ways, that there are other historical options. They must realize that Brazil can forge a different human history, that peripheral countries have an important role to play in producing more stable forms, and that they can and should construct their present and their future in such a way as to contribute to what Santos calls 'another globalization': a globalization that has different uses, where technological and philosophical changes of the human being are complementary; a globalization capable of lending new meaning to each person's existence and also to that of the planet (Santos 2000).

Milton Santos's theories, systemic vision, categories of analysis, concepts and notions, some of which have been used here, encourage an inward movement of reflection, both theoretically/methodologically and politically, about the current stage of technical-scientific-informational development. He has developed the 'right lenses for interpreting the contemporary world' (Silveira 1996: 10), bequeathing to us an invaluable project.

References

Bernardes, J. A. (2007). Dimensões da ação e novas territorialidades no cerrado brasileiro: Pistas para uma análise teórica. *Revista NERA, 10*(10), 1–10.

Bernardes, J. A. (2011). O novo tempo do capital globalizante e as novas relações campo-cidade. In M. A. Saquet, J. C. Suzuki & G. J. Marafon (Eds.), *Territorialidades e diversidade nos campos e nas cidades latino-americanas e francesas* (pp. 195–207). São Paulo: Outras Expressões.

Delgado, G. C. (2012). *Do capital financeiro na agricultura à economia do agronegócio.* Porto Alegre: UFRGS Editora.

Martins, J. S. (1997). *Fronteira: A degradação do outro nos confins do humano.* São Paulo: Contexto.

Polany, K. (2000). *A grande transformação: As origens da nossa época.* Rio de Janeiro: Elsevier.

Ribeiro, A. C. T. (2002). Paradigmas e tendências nos estudos urbano-regionais contemporâneos. *Anais do IV Colóquio sobre transformações territoriais.* Montevidéu.

Ribeiro, A. C. T. (2009). Prefácio. In J. A. Bernardes & R. C. Arruzo (Eds.), *Novas fronteiras da técnica no Vale do Araguaia.* Rio de Janeiro: Arquimedes Edições.

Sánchez, J. E. (1995). Harmonious development or exclusion from productive circuits? *Simpósio internacional: Desenvolvimento sustentável e a geografia política.* Rio de Janeiro: IGU/UGI/LAGET.

Santos, M. (1980). The devil's totality. *Antipode, 12*(3), 41–46.

Santos, M. (1994). *Técnica, espaço e tempo: Globalização e meio técnico-científico-informacional.* São Paulo: Hucitec.

Santos, M. (1996). *A Natureza do espaço: Técnica e tempo; Razão e emoção.* São Paulo: Hucitec.

Santos, M. (2000). *Por uma outra globalização.* Rio de Janeiro: Record.

Santos, M. (2003). *Economia espacial: Críticas e alternativas.* São Paulo: Edusp.

Santos, M., & Silveira, M. L. (2001). *O Brasil: Território e sociedade no início do século XXI.* Rio de Janeiro: Record.

Silveira, M. L. (1996). Milton Santos: uma obra, uma teoria. *AGB Informa, 62*, 10–11.

Chapter 10
Geography and Indigenous Peoples: Milton Santos and the Richness of the Present Time

Roberta Carvalho Arruzzo

Abstract Brazil is a country of significant sociocultural diversity where multiple, sometimes opposing and often conflicting, world views coexist. Most indigenous peoples now inhabit villages in the almost 700 Indigenous Lands already legally constituted or in the process of being recognized and legalized. One of the main issues for these peoples today is their struggle for official recognition of their traditional territories. Most of the Indigenous Lands that have already been demarcated are in the north of the country, especially the Amazon. However, there are indigenous groups living on far smaller lands in every other region of the country where serious territorial conflicts occur. Importantly, the survival of indigenous people in high-value areas is still jeopardized even after their lands are legalized in the form of Indigenous Lands. When their lands are limited to small areas surrounded by deforested land and single-crop farming, their ability to pursue traditional economic activities is curtailed. This chapter explores the work of Milton Santos—especially his concepts of geographical space, used territory, slow people and opaque spaces—with respect to current conflicts over Indigenous Lands. The profound relationship the author establishes between the different actors and technical systems enables an in-depth understanding of the plurality of ways space is constructed in Brazil today. His ideas also allow us to perceive and recognize the wealth of potential ways of living in the present time. In the contemporary moment it is the slow people who can create promising alternatives to the perverseness of globalization and the single discourse prevailing in the world.

Roberta Carvalho Arruzzo, Assistant Professor, Department of Geography, Rural Federal University of Rio de Janeiro, Brazil; Email: roberta.arruzzo@pq.cnpq.br.

© Springer International Publishing AG 2017   113
L. Melgaço and C. Prouse (eds.), *Milton Santos: A Pioneer in Critical Geography from the Global South*, Pioneers in Arts, Humanities, Science, Engineering, Practice 11, DOI 10.1007/978-3-319-53826-6_10

10.1   Introduction

Brazil is a multiethnic and multicultural country.[1] The apparent primacy of Brazilian Portuguese conceals a diversity of over 150[2] languages spoken on a daily basis. Much of this rich linguistic reality can be attributed to the presence of at least 241[3] different indigenous[4] peoples with diverse cultures, histories and ways of living and being. According to the most recent census conducted by the Instituto Brasileiro de Geografia e Estatística (IBGE), there were 896,917 indigenous individuals in Brazil in 2010; approximately 36% live in towns or cities and 64% live in rural areas.[5] Over half of existing indigenous groups live in the almost 700 Indigenous Lands that currently exist in Brazil.[6] Little by little, Brazilian geographers are beginning (albeit tentatively) to understand that this extraordinary ethnic plurality is worthy of investigation: this plurality offers a profusion of ways to think of and operate in geographical space and points to the ways of being that are at stake in contemporary territorial conflicts.

I believe that a theoretical and conceptual approach sensitive to geography will enrich the debate and analysis of Brazil's indigenous issues, specifically with respect to indigenous peoples' rights to territory and to the regularization of their lands. It will, importantly, tease out the fundamentally territorial implications of current conflicts. Milton Santos's thoughts on geographical space, especially his

---

[1]Although the Brazilian ethnic spectrum is quite diverse, in this text I address the originary ethnic groups, also called indigenous peoples, because they have a very specific history that is rarely addressed by Brazilian geography. The term ethnic group (Barth 1998) is used as an alternative to genetic and phenotypic views of socio-cultural differences.

[2]According to Instituto Socioambiental (ISA), data from 2014. See: http://pib.socioambiental.org/pt/c/no-brasil-atual/linguas/introducao. This figure could be even higher. According to the latest Instituto Brasileiro de Geografia e Estatística (IBGE) census, there were 274 languages spoken and 305 indigenous peoples in Brazil in 2010. However, as IBGE itself recognizes, 'when it comes to the total number of languages and ethnic groups, there is still a need for more in-depth linguistic and anthropological studies, since some declared languages could be variants of the same language, while some ethnic groups could also constitute subgroups or segments of the same ethnic group.' See the document 'O Brasil Indígena' in Portuguese at: http://www.funai.gov.br/index.php/indios-no-brasil/o-brasil-indigena-ibge. For further information, see: http://indigenas.ibge.gov.br/graficos-e-tabelas-2.

[3]Instituto Socioambiental (ISA), data from 2014. See: http://pib.socioambiental.org/pt/c/0/1/2/populacao-indigena-no-brasil.

[4]I am considering as indigenous here all those that recognize themselves or are recognized by other indigenous as such. Indigenous communities are based on kinship of neighbourhood relationships and have historical-cultural links with pre-Colombian indigenous societies. For more please see: https://pib.socioambiental.org/pt/c/no-brasil-atual/quem-sao/quem-e-indio and https://pib.socioambiental.org/files/file/PIB_institucional/No_Brasil_todo_mundo_%C3%A9_%C3%ADndio.pdf.

[5]See: http://www.funai.gov.br/index.php/indios-no-brasil/o-brasil-indigena-ibge.

[6]According to the 2010 IBGE census, 57.5% of the indigenous population today lives in officially recognized Indigenous Lands. See 'O Brasil Indígena,' based on the 2010 census, at: http://www.funai.gov.br/index.php/indios-no-brasil/o-brasil-indigena-ibge.

thought-provoking concepts of opaque space and slow people, offer a potentially productive way of thinking through indigenous struggles in Brazil today. Although Santos did not conduct research with indigenous peoples directly, his thoughts can help elucidate the conflicts that these peoples face. In developing the approach, I first contextualize the spatial circumstances of Brazil's indigenous peoples. I then discuss Santos's concepts that help us better understand the spatiality of indigenous peoples in Brazil. In particular, I focus on the relationship he establishes between different actors and technological systems; Santos's approach enables us to probe the multiple ways in which space is constructed in Brazil today, and to understand and value the abundance of different ways of living at the present moment. Ultimately, slow people, as described by Santos, can serve as wellsprings for the creation of equitable alternatives to the perverseness of globalization and its hegemonic 'single world' discourse.

10.2 Territory and Indigenous Peoples in Brazil

Great swathes of land in what is now Brazil are often identified in public policies as great empty spaces. These so-called empty spaces (defined as such in economic or population terms) have been the targets of government discourse and action. However, these spaces were not, historically, empty. Portuguese colonizers[7] and others that embarked on expeditions for minerals, resources and labour found these spaces divided up amongst many indigenous ethnic groups who had pre-existing social arrangements amongst themselves. The relationships and conflicts that arose from conquest were myriad. Oftentimes, indigenous people were dichotomously dubbed either 'gentle indians' or 'fierce indians' in a bid to classify their relationships within colonial categorization frameworks.

These historical relationships continue to impact Brazil's indigenous ethnic groups. As Chauí (2000) describes, a ramification of violent colonization and the construction of Brazil is the prevailing belief that the peoples who lived here previously are of the past. This past is understood in three different ways: *chronologically*, in which indigenous peoples are seen as the remnants or remains of an inevitable process of extinction, which happened through genocide and integration into the national society; *ideologically*, in which these peoples did not keep pace with the progress of civilization and were thus left behind; and *symbolically*, in which they survive only as a memory of the good savage and the harmonious relationship between people and their environment that once existed in these lands.

---

[7]Although here I am making reference to Portuguese colonization, similar actions have since been replicated in myriad initiatives, including public and private colonization projects of the 20th century.

Try as they might, colonial processes have not erased indigenous peoples. Population data indicate that the number of indigenous people is on the rise[8] in recent decades. According to IBGE,[9] the self-declared indigenous population was 294,131 in 1991, rising to 734,127 in 2000, and 817,963 in 2010.[10] The geographical spread of indigenous groups across every region of Brazil means that contact with non-indigenous peoples occurs under very diverse circumstances. The table on the next page displays basic information on the spatial distribution of indigenous groups, measured by the area designated as Indigenous Land.[11] Most officially demarcated indigenous areas are in the north of the country,[12] occupying almost 30% of the landmass in the region. However, there are indigenous groups in every other region of the country, where they occupy far less extensive lands.

The data in Table 10.1 demonstrates that almost 80% of the country's Indigenous Lands are located in the north, while only 16% of Brazil's total farmland is in this region. If we compare this data with Table 10.2, it is apparent that less than half of Brazil's indigenous population lives in the north—approximately 38%. This means that the majority (62%) of indigenous peoples live in much smaller and more densely inhabited Indigenous Lands in other parts of the country, where the majority of farming activities are also conducted. This conflict over proximate space has triggered serious territorial disputes.

Indigenous Lands are a legally constituted category based on the national constitution of 1988. They are collective areas for the exclusive use of indigenous peoples. In theory, the designation of Indigenous Land was designed to break down a longstanding legal tradition of treating indigenous groups as an element of the past. Historically, Brazil's legal framework considered indigenous peoples as groups that survived and needed protection—indeed, should be protected—but who were destined to mix with 'civilized' society and thus lose their legal right to the lands they occupied. This loss of rights was progressively accomplished through, for example, the demarcation of very small indigenous reserves,[13] to which groups from different regions were transferred often by force, thus making their territories available for other activities. The purpose of these reserves was to create areas

---

[8]Although the population of many indigenous ethnic groups is rising, some have a very low population and run the risk of dying out. According to Instituto Socioambiental, seven groups currently have a population of between five and 40 individuals. See: http://pib.socioambiental.org/pt/c/0/1/2/populacao-indigena-no-brasil.

[9]Instituto Brasileiro de Geografia e Estatística (IBGE), which since 1991 has included indigenous peoples in the national demographic census.

[10]In the demographic census of 2010 (IBGE 2012).

[11]In this case, IBGE only counted officially demarcated Indigenous Lands.

[12]This includes most of the Brazilian Amazon.

[13]For example, indigenous reserves were created during the operation of the first state-owned and non-religious agency of support to indigenous peoples, the Serviço de Proteção aos Índios e Localização dos Trabalhadores Nacionais (SPILTN, which became SPI in 1918). The agency was created in 1910 and had as one of its goals the 'civilization' of indigenous peoples. Amidst strong criticisms about mismanagement and corruption, the agency was closed in 1967 and gave way to the Fundação Nacional do Índio (FUNAI) which continues to operate in Brazil.

Table 10.1 Areas of land designated as farmland and Indigenous Land, per region, 2006. *Source* IBGE, Farming Census, 2006

Brazil and regions	Area (hectares)				
	Area of land	Farmland (agriculture and livestock)		Indigenous Land (officially demarcated)	
		Total	%[a]	Total	%[a]
Brazil	851,487,659	329,941,393	38.74	125,545,870	14.74
North	385,332,720	54,787,297	14.21	100,419,452	26.06
Northeast	155,425,696	75,594,346	48.53	2,914,584	1.87
South	57,640,956	41,526,148	72.04	343,283	0.59
Southeast	92,451,127	54,236,169	58.66	128,537	0.13
Central west	160,637,148	103,797,329	64.61	21,740,014	13.53

[a]Percentage of the area of the region

Table 10.2 Self-declared indigenous population according to the 2010 IBGE census. *Source* IBGE census, 2010

Regions	Number of indigenous individuals (self-declared)
North	342,836
Northeast	232,739
South	78,773
Southeast	99,137
Central west	143,432

where indigenous peoples from different ethnic groups could live together while they progressively integrated into the national society as workers. The constitution of 1988 represented a distinct shift in the state's approach: indigenous peoples, their right to land, and their right to maintain their own way of life and culture ceased to be understood as transitory.

For the first time, entire articles in the Brazilian constitution were devoted to indigenous societies; indigenous rights are specifically addressed in articles 231 and 232. These articles ensure indigenous peoples the right to maintain their 'social organization, customs, languages, beliefs and traditions,' (Brasil 1988) as well as the right to occupy the lands they have 'traditionally' occupied. This latter issue is important because it recognizes the necessity of peoples' cultural reproduction through access to specific ancestral lands, rather than just their economic survival through access to any unused land. The articles thus make provisions for 'lands regarded as sacred, distant graveyards, and space for roaming' (Brasil 1996: 12).

Indigenous peoples' ownership and use of the land is assured, in theory, in perpetuity. Third parties can exploit the natural resources found on these lands only with the authorization of both the National Congress and the community in question. Approval to remove indigenous peoples from the lands must be secured from the National Congress, and is only allowed in extraordinary circumstances such as during natural disasters or epidemics. These measures, alongside the possibility of indigenous groups gaining increasing legal autonomy, have ultimately made more secure the maintenance of indigenous cultures, ways of life and rights to traditional territory.

However, the legalization of Indigenous Lands in Brazil is still far from complete. Only 63% of the 679[14] Indigenous Lands have been fully legalized[15], and 19% are still 'under study'. To be 'under study' means that they are still at the first stage of the process, when reports are written to justify the identification and delimitation of certain lands.[16] The process of identifying, delimiting, demarcating and legalizing Indigenous Lands is slow, and reduced further by attempts to amend the legislation[17] to further waylay the legalization process. Different reactions to the demarcation and legalization of Indigenous Lands have been voiced locally, regionally and nationally, in which stereotyped arguments prevail, such as that there is 'plenty of land for so few indians'.[18]

Likewise, the difficulties faced by indigenous groups do not disappear just because their legal ownership to the land has been established. This is particularly the case on lands whose value has appreciated significantly. Many of these lands are trespassed by farmers. The trespassing creates constant conflict with indigenous peoples, and even places them in great peril.[19] The demarcation of a specific land 'liberates' adjacent areas for business interests, putting at great risk traditional activities on Indigenous Land. For instance, Indigenous Lands are often limited to small areas that effectively become marooned in the midst of deforestation and monocultural agriculture. Consequently, traditional activities like hunting and farming are jeopardized; the large areas of adjacent land stripped of their forests not only restricts the area available for hunting, but also modifies the drainage system.

---

[14]See FUNAI: http://www.funai.gov.br/index.php/indios-no-brasil/terras-indigenas.

[15]There are lands that are not fully legalized in every Brazilian region, with great disparities existing between states. Mato Grosso is one of the most severe cases with more than 28% of the Indigenous Lands still to be regularized.

[16]In these cases, the lands have not been delimited and there are no legal assurances. This is the case for many Guarani and Kaiowá lands, where serious armed conflict has resulted in the death of leaders due, in part, to the slow pace of this process. I return to this case later in the text. See: http://www.survivalinternational.org/tribes/guarani e http://www.guarani.roguata.com/text.

[17]The proposed constitutional amendment PEC 215 would transfer the legalization of protected areas (Indigenous Lands, maroon communities and conservation areas) from the Executive to the Legislature. If this does not halt the legalization process, it will at least make it far slower. For more on this matter, see: http://agenciabrasil.ebc.com.br/en/direitos-humanos/noticia/2014-12/vote-pec-215-put-after-protests.

[18]For more, see: http://www.socioambiental.org/pt-br/blog/blog-do-isa/muita-terra-para-pouco-fazendeiro.

[19]See, for instance, the case of the Yanomami http://www.survivalinternational.org/tribes/yanomami.

10.3 Slow People and Opaque Spaces: Proposing a Geographical Perspective on Indigenous Land Issues

As I have shown above, there are many territorial challenges faced by indigenous peoples. Because these challenges are about territory, it is vital to understand them from a geographical perspective. I believe that some of the theoretical concepts developed by Milton Santos are invaluable for understanding these issues. Without his approach, we risk continuing to obscure the current conflicts over Indigenous Lands, thus maintaining and entrenching the prejudices that many indigenous groups face in Brazil today. While most Brazilian geographical studies addressing Indigenous Lands draw primarily on the concept of territory (for example, see Guerra 2011; Motta 2013; Mondardo 2014), I propose that we enhance their concept of geographical space through the work of Santos. In particular, his notions of opaque space and slow people assist in elucidating the factors that underlie territorial disputes. What typical studies of territory ignore, and what Santos allows us to see, is how these conflicts are rooted in, amongst other things, diverse and oft-times opposing worldviews and ways of operating in space.

The concept of space is central to Santos's work. He defines it as the 'indissociable set of systems of natural or manufactured objects and systems of actions, deliberate or not. In each time, new objects and new actions join the others, modifying the whole both formally and substantially' (Santos 1996: 49). Thus, space is materiality and action, synchronicity and asynchrony.[20] Although everything in space is in the present moment, different forms have quite varied temporal origins. As such, the 'simultaneity of different times on a piece of the Earth's crust is what constitutes the specific domain of geography' (Santos 1999: 127). The objective of geography, then, is to study the simultaneity of myriad temporalities that often go unnoticed or are even intentionally undervalued, because 'space is what unites us all, with its multiple possibilities, which are different possibilities for the use of space (territory) related to different possibilities for the use of time' (Santos 1999: 127). The hegemonic and fluid uses of space tend to erase a large range of other uses that are not only possible, but already exist.

This coexistence of multiple times and spaces is often confusing and conflictual, but there is one time and action that is hegemonic, globalizing and seeks to create ever more fluid spaces,[21] that is to say, spaces where actors, merchandises and information circulate with great ease. To obtain this fluidity, hegemonic actors

---

[20]'Since space never contains technologies of the same age or of synchronous variables, one could say that it is an asynchronous space that at once reveals and organizes synchrony. The elements of space, when considered as part of a concrete whole, a place, are seen as synchronous' (Santos 1996: 66).

[21]Santos also draws attention to the fact that 'fluidity is a condition, but hegemonic action is based on competition' (1996: 34), which, considering how it is manifested, 'does not need any ethical justification, indeed, like any other form of violence' (1996: 35).

increasingly imbue this space with science, technology and information,[22] but these elements are not distributed uniformly across the planet. Thus, some areas are more intensely influenced by science, technology and information. These are, according to Santos, luminous spaces. Other zones in which these elements are almost completely absent are called opaque spaces.[23]

Cities are places of simultaneity. They consist of both luminous spaces and opaque spaces, the latter which are focal points for resistance. While urban areas may be modernized, they contain many spaces of opposition 'where time is slow, adapted to infrastructure from the past, where opaque spaces take shape as zones of resistance. It is in these spaces, constituted of non-current forms, that the non-hegemonic economy and the hegemonized social classes find what they need for survival' (Santos 1996: 79). Opaque spaces, inhabited by slow people, are the 'spaces for approximation and not (like luminous zones) spaces for precision; they are inorganic, open spaces and not rationalized and rationalizing spaces, they are spaces for slowness and not for vertigo' (Santos 1996: 83).

Hegemonic discourses and actions value fluidity, and a slow person 'faces up to the strongest manifestations of [this] dominant ideology, such as those related to speed and effectiveness, and enables the many other [slow people] to be valued (and learned from)' (Ribeiro 2012: 60). Valuing other temporalities and ways of being resists attempts to 'flatten' space, to reduce its complexity through actions, understandings and discourses that 'wipe memories, knowledge, projects, and meanings of action, and nullify the achievements of the slow person' (Ribeiro 2012: 66). Hegemonizing processes seek to create 'one world' (Santos 1996: 35) and 'a single world discourse' (Santos 2000). Space can offer a means to resist: 'space appears as a substrate that welcomes the new, but resists change, holding onto the vigor of material and cultural heritage, the strength of what is created from within and resists, the tranquil strength that waits, watchful, for the time and chance to rise up' (Santos 1996: 37). The single discourse of the world only exists as a fable, a trap. Those caught up in its triumphal speed are unable to perceive the illusions of this discourse because they are too enrolled in it. Yet 'the slow person, for whom such images are mirages, cannot for long keep step with this perverse imaginary, and ends up discovering the fabulations' (Santos 1999: 261). This is why the 'strength is with the 'slow'' (Santos 1999: 260) and why the city-dwelling poor, who live in opaque zones, are the ones who 'stare hardest into the future' (Santos 1999: 261).

These processes do not only take place in cities. For Santos, the modernizing countryside is more vulnerable to transformation due to the demands of a hegemonic, externally imposed rationale, precisely because the countryside's materiality

---

[22]Constituting what the author calls the techno-scientific and informational milieu (Santos 1996, 1999, 2000).

[23]There also exist an 'infinity of intermediate situations' (Santos 1996: 52).

and forms from the past are less dense.[24] It is in rural areas that the dominant rationality spreads more easily and takes root more forcefully.[25] Although Santos believes that spaces for creativity and resistance are more often found in cities,[26] I argue that the Brazilian countryside is also an important locus for resistance to a perverse, hegemonizing rationale.

Santos guides us toward understanding 'slow people' as those who are capable of countering the agricultural modernization processes that all too often hem them in:

> This [hegemonic] rationality presupposes counter-rationalities. These counter-rationalities are located, from a geographical point of view, in the least 'modern' areas, and from a social point of view, in minorities. Minorities are defined by their incapacity to submit completely to the hegemonic rationalities. Ethnic, sexual (gender), and other minorities have more trouble accepting and meeting the demands of the rationality, just as their poor are also better defended because they are more hostile to the traps of consumerism. (Santos 1996: 108)

Indigenous people have an important role in maintaining the 'bastions of counter-rationalities' in the Brazilian countryside.[27] When they are assured of the right to use their lands, which legally belong to the Union and cannot be sold, they can constitute a powerful 'obstacle'—in some cases, an obstacle covering vast areas—that remains stubbornly opaque.[28] The main face of contemporary globalizing processes in the Brazilian countryside is export-oriented monoculture, which is constantly having its cropland turned over to Indigenous Lands, and cannot legally make use of these newly demarcated areas. I therefore propose understanding Indigenous Lands as opaque zones in the Brazilian countryside, zones that cannot be 'flattened out', that materially demonstrate the existence of alternative, resistant rationalities, where the material and cultural forms that are not valued by the prevailing rationality continue to

---

[24]'The city is no longer the locus par excellence of new capital. The locus where hegemonic capital spreads easily is the countryside, where horizontal relationships that are established are based materially on science, technology, and information. Cities are where this rapid, easy spread of new capital is refuted' (Santos 1996: 95).

[25]'The perverse rationality takes root more forcefully in the countryside, especially this subtle rationality that comes in the heart of labour and in the form of a discourse whose intent we do not always understand' (Santos 1996: 96).

[26]'Cities have more pockets of counter-rationality and counter-purpose than the countryside' (Santos 1996: 108).

[27]There are many other groups and so-called 'traditional populations' in the Brazilian countryside who also act as important 'bastions of counter-rationalities', such as the *quilombolas, riparian, caiçaras* and many others. In addition to these traditional populations there are also several other peasant groups, some of whom are linked to social movements that struggle for agrarian reform, such as the Movimento dos Sem Terra (MST).

[28]The existence of such opaque spaces is a burden to some hegemonic actors, particularly landowners. The attempts to modify the constitution and the legislation pertaining to Indigenous Lands are not unfounded, once again bringing to mind the proposed constitutional amendment (PEC 215) mentioned in an earlier footnote. Moreover, there are many rural landowners in the Brazilian legislature (see, for instance, Castilho 2012; Moreira 2012).

exist. As Ribeiro (2012: 68) notes, 'opaque space instates the enigma of the invisibility of the highly visible'.

These encounters between the dominant rationality and the counter-rationalities in the Brazilian countryside are very diverse. Indigenous peoples are a primary example of the many 'slow people' and their resistances that operate within 'opaque spaces.' The cases of the Guarani and Kaiowá represent one of the most serious situations in Brazil today. These groups are in contact with non-indigenous populations that have invaded their territory for more than a century, and have suffered a particularly intensive process of territorial expropriation since the 1970s. This expropriation is related to monocultural activities of modern agriculture such as the production of soybeans, corn and sugarcane. Many family groups have fought this process, carrying out various actions to re-acquire their territories. Major actions include *entradas*, which are efforts to reoccupy (and/or remain in) the territories from which they were or are being expelled.

The (re)entry of the Guarani and Kaiowá groups in these areas can be interpreted in antagonistic ways, demonstrating different ways of understanding space. On the one hand, farmers often interpret these actions as a violent disregard for private property. To the Guarani and Kaiowá, however, they represent the resistance of their ways of being. Resistance is expressed in the very name used by these groups to refer to the territory: *tekoha*. The notion of tekoha encompasses both a way of life and a place, and can be understood as the place where you can live your worldview. According to the Guarani and Kaiowá, they belong to a tekoha and not the other way around. Because these indigenous groups belong to a particular territory through tekoha, many family groups that were transferred to official reservations decades ago eventually return and reoccupy small portions of their territories, even when these lands have since been transformed for agricultural production.

These entrada actions, which have been taking place for nearly four decades, pressure the state to demarcate and regularize new Indigenous Lands, especially since the constitution of 1988. Actions such as entradas are an opposition to the discourse that it is only migrants from other Brazilian states, and the economic activities carried out by them, who have 'occupied' these spaces. They thus throw into question the seemingly naturalized existence of agricultural production in the region. Of course, there have been violent reactions to these indigenous groups attempting to inhabit their traditional territories.[29] These peoples' resistance to expulsion shows, however, the different possibilities for living in Brazilian rural areas, and disrupts the fable of an inexorable hegemonic destination for this land, namely monocultural production. Their resistance demonstrates the strength of the 'slow people.'

---

[29]There are several cases of armed attacks and repossessions involving areas of resettlement, including those that are already in an advanced process of regularization. For an example, see: http://www.bbc.com/news/world-latin-america-34166666.

10.4 Conclusion: Transforming Geographic Thought

Ribeiro (2012) draws our attention to the coloniality (Quijano 2005) present in academic thinking, preventing us from taking bold conceptual and methodological steps and inhibiting us from seeing the wealth of the present time. This 'blindness,' which Santos (2002) calls 'contraction of the present,' treats anything regarded as 'a non-credible alternative to what exists' (2002: 246) as non-existent. Coloniality in our thinking shapes our understanding of certain peoples and situations as residual, ignorant, inferior, local and/or unproductive. Thus, drawing out the idea of slow people and opaque space is a way of valuing and engaging with subjects, actions and spaces, as Santos suggests, where creativity and the power for change prevail.

Often treated as insignificant, of the past, or destined to die out, indigenous peoples and Indigenous Lands still resist. Indigenous peoples are slow people who inhabit opaque zones in the Brazilian countryside as well as in cities. Their lives and ways of being comprise the synchronicity of space; they offer strength, resistance, and inventiveness in being able to perceive the fables of the 'single world discourse.' By so doing, they make their strength as political subjects increasingly visible. Indeed, the present time is rich, and this richness is made evident through a sophisticated understanding of space, its opacities, its hubs of resistance and creativity. To think this way inspires a much-needed, fundamental dialogue with the many slow people who live in the countryside. Santos moves us toward this goal.

References

Barth, F. (1998). *Ethnic groups and boundaries: The social organization of culture difference.* Long Grove, IL: Waveland Press.

Brasil, Presidência da República—Casa Civil. (1988). *Constituição da República Federativa do Brasil. Texto constitucional de 1988,* series. Retrieved August 20, 2016, from http://www.planalto.gov.br/ccivil_03/constituicao/Constituicao.htm.

Brasil. (1996). *Sociedades indígenas e a ação do governo.* Brasília: Presidência da República. Retrieved August 20, 2016, from http://www.biblioteca.presidencia.gov.br/publicacoes-oficiais/catalogo/fhc/sociedades-indigenas-e-a-acao-do-governo-1996.pdf.

Castilho, A. L. (2012). *Partido da terra: Como os políticos conquistaram o território brasileiro.* São Paulo: Contexto.

Chauí, M. S. (2000). 500 Anos-caminhos da memória, trilhas do futuro. In L. D. B. Grupioni (Ed.), *Índios no Brasil.* São Paulo: Global.

Guerra, E. F. (2011). Gestão territorial na terra indígena Xakriabá e a geopolítica das retomadas. *Revista Geográfica de América Central, 2*(47E), 1–16.

Mondardo, M. L. (2014). A geometria de poder do conflito territorial entre fazendeiros e Guaranis-Kaiowás na fronteira do Brasil com o Paraguai. *Acta Geográfica* (UFRR), 185–202.

Moreira, R. (2012). *Formação espacial brasileira: Uma contribuição crítica à geografia do Brasil.* Rio de Janeiro: Consequência.

Mota, J. G. B. (2013). Movimento étnico-socioterritorial guarani e kaiowa no estado de mato grosso do sul: Disputas territoriais nas retomadas pelo tekoha-tekoharã. *Revista Nera, 21,* 114–134.

Quijano, A. (2005). Colonialidade do poder, eurocentrismo e américa latina. In E. Lander (Ed.), *A colonialidade do saber: Eurocentrismo e ciências sociais. Perspectivas latinoamericanas.* Buenos Aires: CLACSO. Retrieved August 20, 2016, from http://bibliotecavirtual.clacso.org.ar/.

Ribeiro, A. C. T. (2012). Homens lentos, opacidades e rugosidades. *Redobra, 3,* 58–71.

Santos, B. S. (2002). Para uma sociologia das ausências e uma sociologia das emergências. *Revista Crítica de Ciências Sociais, 63,* 237–280.

Santos, M. (1996). *Técnica, espaço, tempo: Globalização e meio técnico-científico-informacional.* São Paulo: Hucitec.

Santos, M. (1999). *A Natureza do espaço: Técnica e tempo; Razão e emoção.* São Paulo: Hucitec.

Santos, M. (2000). *Por uma outra globalização: Do pensamento único à consciência universal.* Rio de Janeiro: Record.

Chapter 11
Milton Santos's Thought and the Logic of Environmental Conservation in the Contemporary Period

Maria Tereza Duarte Paes and Claudia Levy

Abstract The notion of technique enables us to consider the geographic space as an inextricable system of objects and actions, of instrumental and social means. Reality is historically constituted through this interplay of nature and society. We draw on Milton Santos's notion of technical phenomenon to comprehend the logic of environmental conservation areas in the contemporary period. These new reserve territories designated as conservation units to safeguard heritage are objects of specialized technical knowledge, whereby science becomes subservient to a global market logic, structuring a new system of nature. Biodiversity reserves thus become reserves of value, with subsequent place-based effects. Our empirical research in Brazil links the local to the global scale, while reflecting on how such a verticalized intervention limits important socio-spatial practices and livelihood activities. Establishment of conservation areas involves a negotiation over a given territory's uses whereby hegemonic actors and managerial demands destabilize the normative base of existence of pre-existing social groups, which most often lack information and support to access any rights they may have.

Maria Tereza Duarte Paes, Professor, Institute of Geosciences (IG), Department of Geography (DGEO), University of Campinas (UNICAMP), Brazil; Email: tereza.paes@ige.unicamp.br.

Claudia Levy, Ph.D. Candidate, International Center for Development and Decent Work (ICDD), University of Kassel, Germany; Email: claudia.levy@ymail.com.

© Springer International Publishing AG 2017   125
L. Melgaço and C. Prouse (eds.), *Milton Santos: A Pioneer in Critical Geography from the Global South*, Pioneers in Arts, Humanities, Science, Engineering, Practice 11, DOI 10.1007/978-3-319-53826-6_11

11.1   Introduction

'Meaning makers' (Santos 1996: 103) in modern society produce ideologies and symbolic universes of 'development'. They do so intentionally. Within the realm of conservation, sustainable development[1] has emerged as a key ideological process that, in part, provides for the institutionalization of biodiversity reserves. These new reserve territories, normed by specialized technical knowledge, structure a new *system of nature* (Santos 1992). Accordingly, they serve as 'reserve[s] of value for attainment of future capital' (Becker 2005: 74) to the detriment and destruction of historical territorialities and peoples in these regions.

Santos's notion of technical phenomenon contributes to the analysis of the logic of conservation areas in the contemporary period. He theorizes that, 'nature and space redefine themselves through technical evolution. The periodization of this evolution can serve as basis to acknowledge the periodization of territorial history, culminating in the current phase, in which the problem of 'environment' is imposed' (Santos 1995: 697). Santos's concept of technique goes beyond that of technology and is defined as the main form of relationship between humanity and milieu. Techniques are the 'ensemble of instrumental and social means by which humanity realizes life, produces, and creates space' (Santos 1996: 25). This notion is central to both the constitution of reality and our analysis of said reality. Techniques are thus 'integrated into the environment as a unitary reality' (Santos 1996: 35). Santos urges scholars to understand 'technical phenomenon from a philosophical perspective, i.e., as a whole' (Santos 1996: 20), as a totality. Through technique it is possible to break with the dichotomy between the human and the non-human, the natural and the cultural, the objective and the subjective, the global and the local. The notion of technique moves us away from duality, which is the foundation of modern science. Technique, as *systems* rather than *dichotomies* of objects, has different empirical realities in each epoch's history. It is here, in historical empirical realities, that the notion of *technique* meets the geographic milieu. According to Santos (1996), to consider the geographic space as an inextricable system of objects and actions through the notion of technique, means to assume the inextricability of nature and society, which has varied characteristics throughout history.

---

[1]The term 'sustainable development' is influenced by the notion of *ecodevelopment*. It is the basis of the way of life of traditional populations—above all, those in isolated regions—who seek relative balance, or sustainability, between their socio-cultural reproduction and use of natural resources. The rise of conscience about the 'the limits to growth' (Meadows et al. 1972) incited the debate on the planet's socio-environmental unbalance. The term sustainable development became official following the 1987 *Report of the World Commission on Environment and Development: Our Common Future*. From here on, this term has become almost magical, obligatory to any discourse in the areas of ecology, science or politics that aim to be up to date in international debates.

Santos (1996) proposes three major periods for understanding the history of the geographic milieu: the pre-technical, technical and technical-scientific-informational. First, the pre-technical period is characterized by technical systems' absence of autonomous existence. Here social groups lived in direct relation with natural elements and with local logics. To a great extent before the Agrarian Revolution, to survive meant to outlive the environment's hardships. Here, as a rule, uses of natural resources were place-based. Community-based territorial rules safeguarded social and natural reproduction. Examples of these rules include: 'land set-aside, land rotation, shifting agriculture, which are at the same time social rules and territorial rules, prone to conciliate nature's use and 'conservation': to be used again' (Santos 1995: 700).

In the technical period—Santos's second period—an instrumental logic dominates the logics of nature. Here space is mechanized through technical systems, a function of an international division of labour rather than local logistics. This process, already present in some industrial cities in the 19th century, creates environmental problems such as pollution and ill health. At this point in history, Santos (1996: 189) argues that 'anti-mechanisation reactions, carried out by the various Luddisms, anticipat[ed] the current battle of the environmentalists'.

The third period is the technical-scientific-informational milieu. This period sees a convergence of science and technique, where science becomes subservient to a global market logic. It is in this period of scientific and informational objects that the environmental crisis is consolidated, as seen below. Santos describes nature in this period as global and artificialized; it has become an essential resource of competition among territories: 'Thus nature is transformed into a real system of objects and no longer of things, and, ironically, it is the ecologic movement itself that completes the process of denaturalizing nature, by awarding it a value' (Santos 1996: 53). Here Santos refers to objects as embedded technique, carriers of intentionality. Thus, in objectifying nature by giving it a monetary value (such as through ecosystem services), conservation becomes the basis of economic activities within the 'new economy of nature', or 'green economy' (Fatheuer 2014: 16).

This objectification begins to take shape in the 19th century when the principle of humankind's common heritage legally bound peripheral countries' free access to natural wealth. From the second half of the 20th century the limits to growth (see note 1) that these countries should adopt were sanctioned (Labrot 1996). The reconstruction of cities destroyed by war was coupled with a belief in technological progress, an increase in industrialized goods' consumption, the presence of peripheral Fordism, and an expansion of the rationalist model for territorial planning. Critics of this capitalist paradigm started questioning unequal development and the society of abundance, and created the perception that many environmental problems, such as acid rain, greenhouse effect and environmental disasters, were becoming global in scope.

In this framework the protection of endangered biomes worldwide gains strength as a central environmental management strategy. Protection involves the territorial designation of conservation units (CUs). However, the creation of CUs limits important socio-spatial practices and livelihood activities such as extraction, agriculture, hunting and gathering—survival pillars for traditional culture—while encouraging new ones such as tourism. Territories to be designated as CUs are selected either for a preservationist project of complete protection, or for a conservationist one that foresees some degree of use. Hegemonic actors implement mechanisms that may include or exclude local populations, ignoring the local traditional order. Operating as an instrument within the global technical-scientific logic, the new rationale of ownership of natural resources brings conservation and social uses of particular lands into debate and is ultimately what is at stake. This logic, operating under a globalized market, is 'ever more precise and, also, more blind, since obedient to a foreign project' (Santos 1996: 66), tied to the benefit of private enterprises, states and hegemonic classes (Santos 1992).

In this context, Santos's notion of *used territory* (Santos et al. 2000) is foundational for a comprehensive understanding of the logic behind the creation of CUs as reserves of value, a process that has subsequent place-based effects. Used territory refers to how specific territories are historically constituted through a system of norms, objects and actions operating at various scales of power. It draws attention to the active space of production and social reproduction, and thus to the material and symbolic base of human action. It also points to mechanisms that underpin the *horizontalities* and the *verticalities* of a territory. Horizontalities refer to the local order and its set of contiguous objects organized by the territory. These objects and orders are regulated through geographically proximate relationships. Verticalities refer to the global order that organizes a sparse and geographically diverse set of objects into a system, unifying various localities into a vertical interdependency (Santos 1998). Vertically established conservation units institute a system of norms and (re)production, guided by elite actors.

11.2   Conservation Units: Vertically Imposed Scientific and Informational Objects

As introduced above, the establishment of CUs institutionalizes new norms and rules of territory use and ownership. In doing so, it imposes verticalities pertaining to the globalized market nexus, and undermines local customary social practices and the socio-spatial foundations of the territory's horizontalities. On the one hand, the local is always in dialectic relation with the global, accepting, rejecting or accommodating verticalities. On the other, the hierarchical interdependencies of verticalities necessarily introduces the new, the novel, which is foreign to, and thus in tension with, the local. The new can and often does erode the 'capacity to manage the local way of life' (Santos 1996: 226). CUs—the category used worldwide to designate areas for integral protection—are the formal materialization of what Diegues (1993) terms the 'modern myth' of an 'untouchable nature'. This myth prioritizes the conservation of a particular concept of nature that, though managed and formerly inhabited by humans, is believed to be untouched by humanity.

At the heart of CUs such as National Parks is thus a political project of modernity. It is an attempt to rationalize and reduce the complexity of our relationship with the world. Based on technical-scientific arguments establishing areas of reserve, it devalues fundamental relationships between people and their environment in the face of a capitalist rational logic. We argue against the heroic discourse of 'safeguarding biodiversity for future generations' couched in a modernizing logic. We must understand real socio-spatial scales—global and local—and temporal moments—past, present and future—that underpin the production of territory. The institutionalization of CUs is a legitimization process for reserves of value, produced by a global system and implanted in diverse localities at the contemporary moment. Our approach, following from Santos, recognizes other processes and relationships that are legitimate. After all, as we argued above, the notions of ecodevelopment and sustainability are derived from knowledge of traditional populations' customary ways of life. The institutionalization of 'natural heritage', which now ironically excludes these populations, does not concern biodiversity exclusively. It is a system of technical, political and economic values, imbued with scientific legitimacy, that destroys previously-existing relationships between peoples and their environments. The idea of nature, here, reproduces the urban and capitalist life model.

Nature is thus a socio-political and cultural construction. It is one pole of the constructed relationship between itself and society. Nature and society, in this sense, are opposed and complementary. They are opposed, as biodiversity is transformed into an object of appropriation and consumption through the process of 'technification'. These objects are then instrumentalized, as we will further discuss, through the institutionalization process of CUs, occurring in both concrete and symbolic forms. In this sense nature and society are complementary to the extent that social structures relativize nature's local value (Santos 1998), reinventing and conferring upon it values of use, exchange or reserve, 'transforming natural elements into social resources' (Santos 1996: 54).

The Brazilian CUs' institutionalization process is marked by uneven power. On the one hand, urban and industrial society has long devastated important biomes. Now, a de-territorialized global order seeks to balance its growth limits through the preservation of natural reserve stocks. On the other hand, traditional populations' everyday lives are disrupted and peoples are barred from their territory.

Santos (1992) warned against a premature conceptual move from a geographic milieu to an environmental milieu. He issued this warning against the ideological burden and the manipulative power of the media and general discourses on the eminence of an environmental crisis. 'When the environment, as Nature-spectacle, substitutes the historic Nature—place of all people's labour—and when the *cybernetic* or *synthetic* Nature substitutes the analytical Nature of the past, the concealment process of history's meaning reaches its peak. It is also, in this way, that a painful misunderstanding takes place among technical systems, Nature, society, culture and moral' (Santos 1992: 102). Thus, the geographic milieu, through its analytical category of socio-spatial scales and the notion of used territory, lends itself to an understanding of reality as historically constituted. That is, it helps us to understand how technique mediates the interplay between modes of production and symbolic reproduction.

11.3   New Use and Management Norms

The designation of conservation units involves establishing a set of norms regulating access to and use of these areas, which results in disputes and processes across political, juridical, cultural and economic issues. To secure biodiversity reserves necessitates globally-constituted technical tools that 'bring disorder to sub-spaces where they are installed, [whereby any] order that they create is for their own sake' (Santos 1996: 228). Thus, many CUs are established as territorial prostheses[2] (Santos 1996) that subvert local territoriality networks, the latter which are themselves constituted by particular notions of time, space and nature in the horizontal milieu.

For instance, in Brazil, the establishment of the first national parks and the directives for biodiversity conservation have been guided by the natural sciences' concept of wilderness. Forest reserves to protect specific endangered species were declared as far back as the late 1800s. It was, however, with the Itatiaia National Park, created in 1937 between the states of Rio de Janeiro and Minas Gerais that a landmark was set with regards to the institutionalization of CUs in Brazil. It was inspired by the North American Yellowstone National Park, where the preservation approach emphasizes the scenic facade of the environment (Pimbert and Pretty 1997). The approach is thus preservationist and biocentred.

The institutionalization of each CU requires a management plan based on ecosystem studies, ideally including socioeconomic studies and a participatory planning process, for its reserve and buffer zone areas. As we explain further in the next section, this plan constitutes the instrument of mediation (Santos 1995)— establishing norms for intervention, access and control—defining the CU's management directives, categories and permissible practices. However, in the majority of the cases such documents are only partially produced (often excluding the participatory planning process, or tackling it superficially) and the CUs lack resources for the documents' proper implementation.

Thus, prioritizing the biodiversity component, CU management strategies may attempt to undermine or superficially placate local cultures. Toward the former, CU categories of integral protection presuppose an uninhabited area. As a result, local indigenous peoples' practices are ignored, excluded or even criminalized by law enforcement through, for example, prohibiting the collection of herbs, wild fruits

---

[2]This notion refers to a 'territorial configuration, which [is]…a historic production and tends to deny pre-existing nature, substituting it with a fully humanized nature' (Santos 1996: 39). Depending on science and technology, this historic production, or *technosphere* in Santos's terminology, 'frequently establishes itself while translating faraway interests and thus adhering to the local like a prosthesis' (Santos 1996: 172).

and building materials; hunting; and visits to sacred groves. However, other management measures are of a compensatory kind so as to minimize the contradictions between the territory's customary uses and the new norms established through the CU. Some examples of these compensatory measures are resettlement, compensation for expropriated families or agricultural losses, and mechanisms for economically integrating local inhabitants through, for instance, new forms of work.

New economic dynamics thus also restructure labour in these regions. For many CUs, tourism represents a revenue solution. However, the tourist industry demands particular forms of expertise with regards to infrastructure, professional training and technical adaptation. Moreover, another form of dispossession takes place with these new economic activities. The services required by tourists, such as gourmet gastronomy or sports activities, are linked to modern objects and knowledges, which become incorporated into the territory, displacing previously-existing ways of being. As asserted by Santos:

> [T]he introduction of new ways of performing engenders imbalances from which result, on the one hand, the migration of local customary leadership and the break with habits and traditions and, on the other, transformations of forms of relating. The latter, which have been slowly shaped over a long period of time, are suddenly substituted by new ways of relating whose roots are foreign and whose adaptation to the local is essentially mercantile. (Santos 1997: 46)

Displacement could be minimized, for example, by strengthening place-based action, through communicative and cooperative tools. Such measures may even enhance the objectives of biodiversity conservation. To understand how requires an analysis that sheds light on what Santos calls 'successive Nature systems, which simultaneously encompass and are encompassed by humankind's content, including actions; beliefs; desires; the overwhelming reality; and perspectives' (Santos 1992: 95). Thus, our focus lies in the socio-cultural contradictions that are involved in the competing territorial uses wrought by conservation practices.

11.4 Effects of Conservation Units in the Reshaping of Places

Vertical valorization and transformation of the environment create new relationships of ownership and thus new territories. Santos's analytical approach of the technique allows us to grasp the notion of place in relation to time (periodization), to space (local, regional and global), and to society (historical/customary and contemporary). Likewise, the part (in this case, CUs) cannot alone represent totality. And the totality, in its movement, cuts across the parts, integrating these into its logic. In Santos's words (1986: 39): 'The totality—which presupposes a shared movement of structure, of function and of form—is dialectical and concrete'.

Places—as formative parcels of the social totality—are qualitatively and quantitatively mutating. Space in this conceptualization is an abstract category realized through places' concreteness. Social dynamics reproduce themselves in concrete places, creating history. Customary populations, until recently isolated, could limit their perceptions of and interactions with social space to those necessary for their own reproduction. In contemporary society, a total understanding of space is required for peoples' reproduction. The internationalization of markets and cultures, new informational technologies and international divisions of labour have connected all territorial fractions into a totality, which empirically impacts all manner of relations across diverse places. To Santos (1996) humankind has, for the first time under globalized capitalism, the potential to undergo an *empirical universality*. In other words, space, as it is increasingly interconnected at this moment, may allow a global society to fulfil itself as a phenomenon.

New spatial forms, such as CUs, take shape through social intentionality, inciting conflict at all scales (Santos 1997). They require an active subject capable of creating formal and functional change. This intentionality manifests through the establishment of CUs: the strategy to preserve biodiversity in the face of natural resources' scarcity is based on intentions to control, plan and manage territory. Through these intentions the territory is constituted as an object of political-administrative planning and norming; it becomes a functional parcel articulated by flows of activities, populations and spatial reorganization operating from the local to the global (Santos 1997). Such is the case for the CU territory.

Hegemonic actors' intentionality constructs the new territory of the CU as an assumed inhabited space and as an object of ordering. But, of course, they are not the only social relations present in this territory. The bundle of social relations and intentions underpinning contradictory uses of CU space—such as conservation versus subsistence—creates a dialectical movement between environmental protection strategies and natural resource utilization. It is only by understanding used territory as a totality that these movements become apparent. According to Santos et al. (2000: 2), the idea of used territory 'allows for a broad consideration of the totality of causes and effects regarding the socio-territorial process'.

Our empirical research demonstrates that individual conservation areas are specific outcomes of the general processes already outlined. For example, the establishment of the Serra da Capivara National Park (SCNP) in the Brazilian Piauí state acted as a prosthesis inserted in the local territory. Management of the park required community resettlement, border delimitation, and surveillance and restraint of local communities who are no longer able to undertake their productive and customary activities, such as hunting and the use of agricultural areas, communication corridors and sanctuaries. This process triggered a series of grievous conflicts. For instance, households that had not resettled were fined or had their livestock sacrificed for crossing the parks' border. These animals also became easy bait for wild animals that invaded the farms. The new terrritorialization—whereby the local population no longer governed the uses of the resources—created room for external actors to illegally exploit resources (Levy 2006). Even the management tools that continue to seek to integrate local populations into productive and cultural activities within the CU tend to prioritize economic efficiency rather than a responsible and reciprocal inclusion process with these communities. These issues are inciting conflict across Brazil and the globe more generally.

The prostheses organized by global capital and the hegemonic elite generally disrupt specific systems of shared rights over land and other natural resources, such as management through the 'land of commons' (Godoi 1999) model or 'common-pool resource systems' (Ostrom 2002). Naming the park legitimates the dispossessing prosthesis. The resulting estrangement of people from their lands and from the common management of resources not only benefits outsiders, but also becomes common sense to many involved. We have observed similar situations with respect to hunting in SCNP, or to wood exploitation in the CU Ankasa in Ghana (Levy 2009). We are also informed by a decade of research in the Serra do Mar plains and hillsides, on the north coast of São Paulo and south of Rio de Janeiro, a region with preserved Atlantic Forest areas populated by artisan fisher folk (*caiçara*) communities. Since the 1970s the expansion of CUs—and thus of environmental law enforcement—has robbed the *caiçara* culture of its customary basis of economic survival—hunting, extractive collection of plants, fruits and wood, and smallholder agriculture (mainly cassava and banana). A process of *touristic urbanization*[3] (Luchiari 2000) tends to follow the direct or indirect expulsion of these populations. Consequently, real estate speculation and a new urban rationality dominate these lands. Once a new culture integrates with the CU ecosystem, the *caiçara* become marginal people in their own territory, living on low-qualified jobs in the urban and tourist markets.

Thus, territoralizations of CUs involve a negotiation over a given territory's uses. However, all actors are not equal in this negotiation. The system of resource use imposed by hegemonic actors is shaped by managerial demands, while pre-existing social groups, now de-territorialized, lack information and support to access any rights they may have. This process is happening on the continent of Africa as well: it is estimated that '14 million people in Africa alone are said to have been driven off their land by environmental conservation measures (e.g. the designation of parks)' (Kareiva, cited in Fatheuer 2014: 59).

The rationalization of activities under a global order—an 'everyday life imposed from outside, ordered by privileged information, which is secret and powerful' (Santos 1996: 108)—tends to prevail in our contemporary world. With the establishment of CUs, we argue that technical-scientific logic becomes an instrument used for domination and control. It operates through a hierarchy of power that issues management directives via decrees, administrative rules and conventions. This process determines which social practices are allowed and manifests in fairly predictable ways amongst subjects who hold unequal power: the horizontal

---

[3]Though boosting different sectors of the economy, the logic of touristic urbanization is one in which consumption prevails over production, and with it the precarization of working conditions, the intensification of migratory fluxes and the hybridization of culture. Furthermore, real estate speculation ends up selling the value of the estheticized landscapes and seasonal infrastructures prioritize uses for an elite. All of these are factors differentiating this process from industrial urbanization, positioning tourism as a vector of place transformation (Luchiari 2000).

relationships of those who have historically inhabited an area that is now a conservation territory tend to be displaced by the vertically ordered relationships of global capital.

11.5 Final Thoughts

Milton Santos did not specifically address the debate on biodiversity conservation. However, his thinking leads us toward a comprehensive view on territorial uses and management in conservation areas. Established by a dominant global demand using technical-scientific logic as an instrument of power (Santos 1998), a new system of nature introduces norms that govern territories in the name of conservation of a global natural heritage. It does so while excluding local knowledges from the relations established between humanity and nature, and brings to evidence the conflicting and complementary dialectic between totality and part, globe and place. Conservation thus becomes the crystallization of the representation of oft-times elite groups—the natural science-centred environmentalist, and the construction and tourism sectors—to the detriment of other, less organized groups on the margins of the capitalist rationale, such as local indigenous peoples.

If we allow the environmental milieu to substitute the geographic milieu, we lend ourselves to the spectacularization of Nature. Understanding the geographic milieu through Santos's technique allows us to comprehend the social totality of the world. Conceptual rigour is fundamental to this task, along with the ethical imperative of acknowledging place-based knowledges (Castro 2000). Of course, the preservationist and bio-centred models of conservation have useful aspects. But there is much to be done before a large number of conservation areas are created at the expense of long-established peoples and their ways of life. A way forward might be through the use of communicative action approaches for the design of CUs' management tools.

References

Becker, B. K. (2005). Geopolítica da Amazônia. *Estudos Avançados, 19*(53), 71–86.

Castro, E. (2000). Território, biodiversidade e saberes das populações tradicionais. In A. C. Diegues (Ed.), *Etnoconservação: Novos rumos para a proteção da natureza nos trópicos* (pp. 165–182). São Paulo: Hucitec.

Delphim, C. F. M. (2009). O patrimônio natural no Brasil. In P. P. A. Funari, S. Pelegrini & G. Rambelli (Eds.), *Patrimônio cultural e ambiental: Questões legais e conceituais* (pp. 167–186). São Paulo: Annablume/Nepam-Unicamp/Fapesp.

Diegues, A. C. S. (1993). *O mito moderno da natureza intocada: populações tradicionais em unidades de conservação.* NUPAUB/USP, São Paulo: Hucitec.

Fatheuer, T. (2014). *New economy of nature: A critical introduction.* Berlin: Heinrich Böll Foundation.

Godoi, E. P. (1999). *O trabalho da memória: Cotidiano e história no sertão do piauí.* Campinas: Centro de Memoria, Unicamp.

Labrot, V. (1996). L'apport du droit international: Patrimoine commun de l'humanité et patrimoine naturel. In Y. Lamy (Ed.), *l'Alchimie du patrimoine: Discours et politiques, editions de la maison de sciences de l'homme d'aquitaine* (pp. 109–135). France: Talence.

Levy, C. (2006). *Gestão e usos do território: Conflitos e praticas socio-espaciais no Parque Nacional da Serra da Capivara, Piaui, Brasil.* Campinas: Unicamp.

Levy, C. (2009). Participatory forest management in Ghana. In K. Gaesing (Ed.), *Reconciling rural livelihood and biodiversity conservation* (pp. 133–144). SPRING Research Series 52. TU Dortmund.

Luchiari, M. T. D. P. (2000). Urbanização turística: Um novo nexo entre o lugar e o mundo. In C. Serrano, H. T. Bruhns & M. T. D. P. Luchiari (Eds.), *Olhares contemporâneos sobre o turismo* (pp. 105–130). Campinas: Papirus, Coleção Turismo.

Meadows, D. H., et al. (1972). *The limits to growth.* New York: Universe Books.

Melgaço, L. (2013). Security and surveillance in times of globalization: An appraisal of Milton Santos' theory. *International Journal of E-Planning Research, 2*(4)1–12.

Ostrom, E. (2002). Reformulating the commons. *Ambiente & Sociedade, 5*(10), 1–21.

Pimbert, M. P., & Pretty, J. N. (1997). Parks, people and professionals: Putting "participation" into protected area management. *Social Change and Conservation,* 297–330.

Santos, M. (1986). *Pensando o espaço do homem.* São Paulo: Hucitec.

Santos, M. (1992). A redescoberta da natureza. *Estudos Avançados, 6*(14), 95–106.

Santos, M. (1995). A questão do medio ambiente: desafios para a construção de uma perspectiva transdisciplinar. *Anales de Geografía de La Universidad Complutense, Madrid, 15,* 696–706.

Santos, M. (1996). *A Natureza do espaço: Técnica e tempo; Razão e emoção* (4th ed., Vol. 4). São Paulo: EDUSP, Universidade de São Paulo.

Santos, M. (1997). *Espaço e Método* (4th ed.). Sao Paulo: Nobel.

Santos, M. (1998). O Retorno do território. In M. Santos, M. A. A. de Souza, & M. L. Silveira (Eds.), *Território: Globalização e fragmentação* (4th ed., pp. 15–20). São Paulo: Hucitec.

Santos, M. et al. (2000). O Papel ativo da geografia: Um manifesto. *Território, 9,* 103–109.

Chapter 12
Environmental Crisis Through the Theories of Milton Santos

Francisco J. Toro

Abstract Milton Santos, in his epilogue and synthetic work *A Natureza do Espaço* (The Nature of Space), asked a vital question, one rarely raised in official environmental discourse: Should we talk about deterritorialization of environmental crisis? In this chapter I explore how a deep engagement with this question might help us face, in a more effective manner, current and future environmental challenges. This chapter uses several works of the Brazilian geographer to develop an ideal theoretical background to better understand and interpret environmental crisis in the contemporary age. My analysis of environmental crisis will consider four dimensions through which Santos's insights may be structured: ontological, epistemic, technical and ethical. Any approach to human-nature relationships must consider both space and territory not merely as scenario, but as meaningful variables, as Santos did. Thus, I use Santos's notions of space and *milieu* as key concepts of environmental crisis. I attempt to show that Santos's theory enriches the environmental paradigm with new theoretical tools, and places the discipline of geography in a strategic position to respond to and challenge epistemological issues of environmental crisis.

12.1 Introduction[1]

The environmental crisis depicts a new age in the history of humankind, inaugurated with 19th century industrialization. The crisis has been exponentially intensifying since the 1950s, a period recently dubbed the 'Anthropocene'

Francisco J. Toro, Associate Professor, Department of Regional Geographic Analysis and Physical Geography University of Granada, Spain; Email: fjtoro@ugr.es.

[1]A sincere thanks to Prof. Margarita Carretero, Department of English and German Philology of University of Granada (Spain), for her comments and suggestions; to Lucas Melgaço, Department of Criminology of Vrije Universiteit Brussel (Belgium) for his specific reading of Santos's insights and suggestions; and to Carolyn Prouse, Department of Geography of The University of British Columbia for her several readings and reviews.

© Springer International Publishing AG 2017
L. Melgaço and C. Prouse (eds.), *Milton Santos: A Pioneer in Critical Geography from the Global South*, Pioneers in Arts, Humanities, Science, Engineering, Practice 11, DOI 10.1007/978-3-319-53826-6_12

(Crutzen & Stoermer 2000) and characterized by the generalization of electricity and fossil fuels, mostly oil. Environmental crisis in the Anthropocene refers to the global rift between the metabolism of human systems and the incapacity of the biosphere to provide unlimited material resources and absorb diverse sources of waste and heat.[2] This environmental crisis produces 'global environmental change' (Turner et al. 1990; Goudie 2006).

Many scholars point out the hegemonic role played by the Western model of development in precipitating environmental crisis (see, for instance: White 1967; Moncrief 1970; Schumacher 1973; Redclift 1984; Hernández del Águila 1985; Leff 1986; Norgaard 1994; Naredo 2006; Sachs 2010; Toro 2011; Escobar 2015). This model is guided by an epistemic and ontological philosophy that represents and perceives Nature[3] in opposition to human issues. According to these scholars, behind environmental crisis there is a cultural problem closely linked to a specific way of representing, knowing and perceiving the world. Many point to the importance of broadening our understanding of 'cultural' to include not only theories, rationalities and philosophies, but also objects, tools, customs and even built environments, which mediate humans' relations with physical attributes and functions of Nature. Following this approach, any theoretical attempts to understand the roots of environmental crisis should interrogate both technical and material transformations of geographic reality.

On this last point it is worth highlighting the epistemic contributions of Brazilian geographer Milton Santos to geography, in particular, and to social sciences, in general. His thought offers a theoretical background that may assist in the epistemic and cultural challenges posed by our current environmental crisis.[4] Unfortunately,

---

[2]Beyond the scientific verification of many environmental disruptions induced and caused by human activity during the last century, the interpretation of the environmental crisis is not monolithic. Indeed, it depends on particular scientific, political or ideological frameworks, producing a wide range of metanarratives in its etiological analysis. During the 1960s and 1970s, with the emergence of modern environmental concern, diverse factors were variously pointed to as the main driving force of environmental crisis, responding to different metanarratives: population growth, economic model of development, technological advances and scale, affluence and consumerism of rich societies and anthropocentric ethics, among others.

[3]Given the polysemy of the term 'nature', to avoid confusion I will use 'Nature', capitalized, to denote the social construct created in opposition to the human-artificial realm, and as that which existed before the significant environmental transformation instigated by human societies. I use 'nature' to refer to the essence of the things, except in the case of works' tittles (e.g. *The Nature of Space*). The adjective 'natural' or 'nature' will refer to the biophysical condition of the environment, milieu or objects subjected to physics and biological rules and which is connected to the idea of 'Nature'. In the case of the use of nature or Nature in Santos's quotes, I will keep the original term.

[4]I synthetize the 'cultural problem' of environmental crisis through five principles: (1) the relativism and polysemy of the idea of Nature; (2) the hegemony of scientific-technical rationality (linear conception of natural processes) in the usage and exploitation of natural resources; (3) the exclusion of nature from Western discourses of progress and development; (4) the construction of economic science in an isolated universe apart from the biosphere and society; and (5) the promotion of consumerist patterns in the post-industrial era (Toro 2011).

this aspect of Santos's theory is not very well known or referred to in published literature and research; yet, its scope takes environmental issues into account, directly or indirectly. As Melgaço (2013: 2) reveals: 'Santos was in fact a specialist in theory, geographical theory to be more exact. He created a set of articulated, coherent concepts that together form a solid and fruitful body of work' including, as I will discuss, thoughts on the environment.

Santos turned to diverse disciplines and scholars from natural to social sciences in order to bolster his arguments. In fact, one of his main concerns was to provide Geography with a distinct conceptual and theoretical framework while breaking the rigid and seemingly impassable borders between neighbouring disciplines. As such, Santos embodies the cross-disciplinary attitude that is demanded by any scientific and theoretical approach to understanding environmental issues; these issues, of course, traverse disciplinary boundaries.

The thought of Santos is a unique combination of fundamental concepts from geographic theory and from philosophy. Theorizing ontological categories, such as technique, space, time and Nature—all of which characterize and influence the human environment—has two virtues: (i) it makes cross-disciplinary dialogue easier by mobilizing shared concepts of different disciplines; and (ii) it remains valid and reliable over time, as he teases out the most essential aspects of *Homo geographicus* (Sack 1997). Indeed, Santos's ontological categories are immanent realities of any historical period and may help to interpret interactions between human and biophysical structures in the Globalization Age.[5]

The main arguments of Santos's theory and perspectives on ontology are found in his most substantial work, *A Natureza do Espaço* (The Nature of Space) (Melgaço 2013: 10), which is a kind of epilogue and compilation of his main theoretical contributions. But he also wrote a number of essays that provide fundamental insights, offering a critical approach to understanding environmental crisis, scientific knowledge and, above all, environmentalism. These include *A Redescoberta da Natureza* (The Rediscovery of Nature) (Santos 1992) and *A Questão do Meio Ambiente: Desafios para a Construção de uma Perspectiva Transdisciplinar* (The Question of the Environment: Challenges for Building a Cross-Disciplinary Perspective) (Santos 1995).

This essay provides a review of Santos's theory as it is relevant for understanding contemporary environmental crisis. I discuss four aspects that help to synthesize the cultural dimensions of the environmental crisis, aspects that Santos

---

[5]The concept of 'globalization' was originally proposed by the economist Levitt (1983) to identify the progressive standardization of markets as a result of the strategies applied by international companies. Roland Robertson was the first sociologist in 1985 to coin the term 'globalization', inspired by world systems theory of Immanuel Wallerstein (Robertson & Lechner 1985). Later, in 1995, Robertson popularized the term 'glocalization', pointing to the intimate and reciprocal connections between local processes and global trends (Robertson 1995). Amin (1997) called one of his most famous works *Capitalism in the Age of Globalization*, in which he criticized the growing inequalities that the globalization of capital is creating.

refers to explicitly or implicitly: (a) ontological; (b) epistemic; (c) technical; and (d) ethical. Finally, I reflect on the Brazilian geographer's question: 'Should we talk about deterritorialization of the ecological crisis?' (Santos 1996: 170).

12.2   Ontological

Humans' relationship with Nature, understood here as a biophysical-environmental reality, is not unmediated. On the one hand, societies and peoples, according to their cultural contexts, share a preconceived idea of Nature that shapes the way it is subsequently perceived, felt and used. Yet, on the other hand, the environmental conditions of the places we live, the landscapes we see, and the buildings we use, also affect and shape our idea of Nature (Tuan 1974). This reciprocal relationship is often confused, and it can thus be difficult to determine which features of Nature are a function of our cultural understandings, which correspond to environmental conditions, and how these interact to shape our perceptions.

Santos approaches the essence and reality of Nature in the global, hybrid and transformed environment where our civilization currently resides. He bases his considerations in history (Peluso 2013), which leads to his first conclusion: we can solely know Nature in its historical context, that is, *the concept of Nature is relative and even enigmatic.*

Historically, humans have tried to imitate the appearance, features and functions of Nature, designing and creating landscapes to adapt to the most diverse human needs and desires. In terms of production, agricultural practices have domesticated Nature, increasing reproduction of selected plants to extract from them the highest profit. Gardens and parks reflect the creation of aesthetic landscapes imbued with symbolic, political and/or leisure intentions. More recently, environmental concerns have led to a yearning for a symbolic Nature, as its healthiest and purest conditions are increasingly threatened by industrial development. As a result, nature parks have been established to protect the last redoubts of a supposed primitive and sacred Nature. Here, individuals socialized within a 'risk society' can meet Nature in urban environments, but in a simulated way, in *a* Nature that is a product of Arts (Beck & Willms 2003), that is, a cultural representation or artificial imitation of the original. Sustainability-oriented policies have been developed by numerous cities to offer the illusion of a balanced reunion with Nature, which is more ideal than real. In such a context, the idea of Nature is confused and mixed, a hybrid (Latour 1991).

Santos initially confronts the ontological problem of Nature through a Marxist approach (Peluso 2013), differentiating between 'first nature' and 'second nature' (Santos 1996). Both *Natures* do not of necessity correspond to consecutive stages of human evolution, where the first yields to the latter; rather Santos refers to the transition through two ontological categories of a pure Nature and of a transformed

Nature: from the 'natural milieu,[6] as societies interact intimately with biophysical components and functions, to the 'technical-scientific-informational milieu', a situation in which the 'natural nature, where it still exists, tends to retreat, sometimes brutally' (Santos 1996: 160). Tautologically, Santos uses the expression 'natural nature' to refer to the marginal presence of biophysical features within the domain of human technical environments and to describe how Nature has been virtually stripped from our daily landscapes, both urban and rural. He asserts: 'But now, when the natural cedes, the artificial and victorious rationality shows itself as an instrumental Nature…it expresses itself to us as supernatural' (Santos 1992: 96). That is, Nature is something increasingly strange in our daily environments, habits and thoughts, almost a myth or utopia.

Santos believes that an appropriate assessment and analysis of environmental impacts requires establishing a clear distinction between anthropogenic Nature ('second nature') and natural Nature ('first nature'). He reminds us that Nature, as 'ecological environment', is 'the set of territorial complexes that constitute the physical basis of human labour' (Santos 1985: 23), the material sustenance of our progress and reproduction. But he is really admitting that we need to be involved and integrated within the original structure and complexity of Nature, since 'second nature' is strongly dependent on and a part of 'first nature'.

Therefore, for the Brazilian geographer it is not accurate to refer to 'impacts on environment' as though Nature is something external to societies, because 'what nowadays are called environmental damages are really damages to livelihood, that is, the milieu seen in all its integrity' (Santos 1995: 697). It is a concretized-spatial way of understanding the idea of Nature. In terms of conceptualization, Nature is inseparable from *milieu*, but the first is a tributary of the latter.

12.3 Epistemic

One of the greatest epistemic challenges faced by humans is confusing the representation of reality with reality itself. As Morin (1999: 6) said: 'Our systems of ideas (theories, doctrines, ideologies) are subject to error and, in addition, they protect errors and illusions contained in themselves.'

---

[6]'Milieu' is a French word that may be translated into English as 'social environment' and offers an historical contextualization of Human-Nature relationships in space. The humanistic geographer Anne Buttimer differentiates four dimensions or 'constellations' of milieu, to which Geography and geographers have paid attention: 'identity (national, regional or local), order (spatial, structural or administrative), niche (resources, *Raum*, or demographic base or potential) and inventory (information and communication of knowledge about one's world)' (Buttimer 1993: 31). For Buttimer, milieu 'opens up a vast range of potential inquiry and reflections about the practice of geography' (Buttimer 1993: 34). According to Melgaço (2013: 3), Santos 'generally used milieu to refer to space, to the way technique is imbricate with geographic space, while period is related to history, time, and processes'.

In his most famous work entitled *The Structure of Scientific Revolutions*, Thomas Kuhn (1962) introduced the concept of 'normal science' as the legitimate way to understand reality in which a scientific community consolidates a regular procedure, that is, a scientific paradigm. In his formulation, the application of a normal scientific method will lead to a valid and objective knowledge in every discipline. This theory has been largely discussed by both partisans (Popper 1965; Lakatos 1978) and critics (Feyerabend 1975; Morin 1977; Funtowictz & Ravetz 1993) of the scientific method.

It is necessary, as a starting point, to clarify what may be understood as scientific method in order to address the limitations of scientific knowledge and the confusion of representation with reality.

Recently the Spanish physicist Wagensberg (2014) attempted to synthetize the scientific method through three fundamental principles: objectivity, intelligibility and dialectics. These principles lead to a systematic, normalized and useful knowledge; however, this representation of reality is extremely simple, reductive and abstract. Moreover, for some critical thinkers, the scientific method and/or normal science may feed the cultural problem of environmental crisis if they reproduce the patterns of Cartesian and mechanistic views of Nature (Capra 1982; Martínez 1993). The extreme specialization, isolation and fragmentation of knowledge make it difficult to comprehend the complex, non-linear and uncertain ecological and social dynamics. Therefore, Santos advocates for an interdisciplinary scientific approach to understand and face environmental challenges.

Santos takes issue with a notion central to the scientific method: the idea of objectivity. The principle of objectivity is 'to attain the maximum universality of the science vis-à-vis the observer, in other words, the least influence from her particular beliefs, prejudices, or circumstances' (Wagensberg 2014: 336). This principle connects with the aspiration for neutrality and asepsis in scientific research. Santos is concerned with science's approach to the reality under analysis. He asserts that: 'if we do not take into account the manifold perspectives of how reality reveals itself to us, it could lead to the theoretical construction of a blind and confusing whole' (Santos 1995: 696). In fact, normal science cannot absolutely control the complex and unpredictable reality: 'features of reality turn autonomous themselves and require an autonomous and special treatment' (Santos 1995: 696).

Climate change serves as a good example. Many scientists have based their analyses on diverse trends of climate change variables to forecast future scenarios. Thus, climate change only *exists* as a prediction of a more or less probable reality, constructed from a set of statistical parameters. But laypeople do not handle probability and speculation very well; they need certainty. It is much easier to trust and believe sides of radicalization of climate change narrative, which are not pre-cisely *neutral* and do not represent complexity: complete negation, supported by the discourse of an environmental conspiracy; or apocalyptic fate, giving extreme and deliberated credibility to specific scenarios. In this regard, Santos suggests to trust moderately in science and use a complex and integral perspective: 'if it is necessary to face pollution and environmental degradation, we must do so with our eyes wide open, basing ourselves on scientific analysis and not just shouting: Fire!' (Santos

1992: 101). This assertion may seem ambiguous, insofar as he does not specify what kind of scientific analysis we need to trust.

Nowadays, science is somewhat of a religion to lay societies (Naredo 2006). It is thought to be central to human progress, thus occupying a privileged position in the epistemic world. However, science and scientists are shaped by values. Because scientific knowledge is produced by inter-subjectivities, scientists and scholars share common patterns, theories and methods, but they are ultimately motivated by different reasons and have different intentions, be they economic, ethical and/or political. The 'principle of objectivity', or the idea that the scientific method produces objective science, is thus overestimated. Anticipating more recent arguments about the neoliberalization of education, Santos reveals how manipulation and selfish knowledge have materialized in the university, once-considered an independent and autonomous institution: 'In the name of scientism, the pragmatic behaviors and technical reasons, which ride roughshod over the efforts of integral understanding of reality, are imposed and rewarded. In a university of results, the will of being genuinely intellectual is penalized, pushing the best talents to a spasmodic research that is statistically profitable' (Santos 1992: 103–104).

Santos is very critical of reductionist analyses as they contribute to the fragmentation of knowledge into disciplines. For instance, the 'principle of intelligibility' often shapes analyses, creating 'the most compact form of understanding… [that] tends to determine the minimum of a maximum' (Wagensberg 2014: 337). According to Santos, the problem of a fragmented knowledge is seen fairly often with respect to environmental issues. The interactions and feedback between natural and human systems far exceed the reductive disciplinary boundaries that attempt to study them. However, Santos does not invalidate the role of disciplines and specializations; as he states, 'the need for metadisciplines that drive a systematic view of reality does not exclude the specializations; they are still necessary' (Santos 1995: 696). Overall, then, Santos believes that complex and multidimensional environmental issues require holistic and global analytic approaches. He appeals for 'a review of disciplinary theories and practices insofar as [the environmental crisis] demands a comprehensive and all-encompassing analysis' (Santos 1995: 695), which necessarily involves an intimate relationship between specializations and metadisciplines.

Moreover, these epistemic patterns of objectivity and reductivity condition the analysis of crisis in a counterproductive manner. Reductions such as a purely ideological view, a purely economic view or an exclusively present concern 'renew the risk of making a causal chain that leads to the absurd in knowledge production whereby effects eventually precede the causes' (Santos 1995: 702). This confusion affects the way in which many environmental problems are being managed. 'Make-up' and corrective measures (also known as end-of-pipe measures[7]) are

---

[7]For instance, technical solutions for exporting pollution in time and space include: the construction of increasingly deeper outfalls into the sea for sewage; or the system of catalysts in cars for reducing emissions of greenhouse gases.

applied instead of preventive agents,[8] the former which feedback positively on environmental risks and their impact in space and time.

Santos does not necessarily oppose the progress of science, if the old knowledge is a 'prisoner [that resists] change and risks falling into the drift of interpreting the present' (Santos 1995: 695). However, the excessive focus on innovation and progress in science may blind the attempt to understand the causes and roots of environmental problems. For Santos, turning to history is an antidote against a kind of scientific activity that 'ventures to work only in and from the present' (Santos 1995: 696). He suggests a historical de-construction of study objects according to their respective contexts and formational processes. In his work, Santos uses this epistemology for understanding the historical roots of contemporary phenomena: he mobilizes a dialectical approach that attains '"the highest degree of falsifiability available" in pursuit of progress of science' (Wagensberg 2014: 339). He argues for a historical approach to understanding concepts such as Nature, environment, milieu, culture and technology in order to learn lessons from the past for facing today's environmental crisis.

A summary here is useful to understand how Santos resolves his ambiguity regarding the value of science and the scientific method. He denounces certain attitudes and streams in normal science: (a) the current research system that penalizes autonomy and creativity of scholars (likely referring to the disciplines of physics, math and technics) and (b) the production of academic knowledge to obtain short-term results, subjugated to an instrumental and technical rationality. In this regard, Santos agrees with other authors claiming that knowledge is becoming capitalized and increasingly oriented to business aims (Blackmore 2001; Krimsky 2003; Lander 2008; Galcerán 2010). Interpreting Santos, we may conclude that if science were to become epistemologically relativist (Boghossian 2006) or be co-opted by business (Porrit 2000), it could lead to the discrediting of important environmental information, allowing media, political discourse and profit-oriented companies to take advantage of a common but not too reliable knowledge (see Sect. 5 'Ethical').

12.4   Technical

Likely the most recurrent concept in Santos's theory is technique, and it can be used to understand the environmental crisis. According to Melgaço (2013), the concept of technique is an abstract and comprehensive approach and is defined by Santos

---

[8]For example, the introduction of biological methods such as natural predators for controlling plagues in agriculture; or prohibiting industrial and transgenic products in the market, of which there is a certain probability of serious problems for both human and environmental health.

'as the main form of relationship between humanity and milieu' (Santos 1996, in Melgaço 2013: 12). For Santos (1996), and following the insights of Séris (1994), every artificial object in the environment, even the natural ones, might be included among technical objects, if the criteria of its possible use are considered. Thus, technical objects have a potential and instrumental use for any given technical activity. This technical activity—that uses technical objects—includes any way of manipulating the environment for human purposes. As such, technical objects and technical activity have a spatial connotation, and should be studied as part of something broader and more complex, a 'technical' environment. In this regard, Santos views the environmental crisis as a technical problem of our contemporary age: 'interpretation [of the environmental crisis] is not possible without bearing in mind, once more, the typology of technical objects and the motivations for their use in the current historical period' (Santos 1996: 170). For him, 'technique is an environment itself', insofar as 'technical objects need to be analyzed jointly by their environment' (Santos 1996: 22). Understanding the environmental crisis must be done by analyzing it as a historical process: 'nature and space are redefined by technical evolution. Technique would be a pivotal reference in every historical moment' (Santos 1995: 696).

Historical analytic approaches often privilege industrialization to categorize human progress into different stages (that is, pre-industrial, industrial and post-industrial). But, according to Santos, industrialization is merely a feature of a broader reality, that is, of technique. The main contribution of Santos is infusing technique with a spatial dimension, which takes into account its specific geography (Santos 1996). Using 'technique', Santos arrives at a different historical sequence. He presents three sorts of milieu: a natural milieu, a technical milieu, and a technical-scientific-informational milieu.

Santos calls the natural milieu the period where 'technique does not have an autonomous existence' (Santos 1996: 157). Here, humans began a kind of domination by domesticating plants and livestock, but with a low level of intensity and degradation. He explains that technique was in harmony with Nature, as long as it was perfectly integrated without compromising the limitations of natural resources. Moreover, some agrarian techniques (such as fallow, crop rotation and shifting cultivation) enhanced certain natural conditions and restored temporary impoverishments of resources. To Santos, the main motivation or rationality of these societies was 'the preservation and maintenance of livelihood' (Santos 1996: 158).

Divergence between technique and forces of Nature characterizes the technical milieu. If technique was previously integrated with natural processes, now technical objects are fashioned by an instrumental logic that 'challenges natural logics and creates...mixed and hybrid conflicts' (Santos 1996: 158). Thus, technique makes its own rules and shapes new environments, to the point that technical objects 'are no longer an extension of bodies, but extensions of territory, real prosthesis' (Santos 1996: 158). In other words, technique is no longer adapted to a 'human scale', and becomes a sign of a 'progress trap' (Wright 2004; Kingsnorth 2013).[9] Mechanical-instrumental rationality superimposes itself onto Nature's logic and supports an economical rationality, that is, actions motivated by maximizing profit and commodifying goods. The first great environmental impacts were created through the technical milieu with the growth of cities, railroads and factories, which incited protests, as Santos recognizes: 'Luddite uprisings heralded environmentalist struggles' (Santos 1996: 159).

The technical-scientific-informational milieu expresses the highest level of spatial transformation, in which reciprocal connections between local and global are produced everywhere. Understanding this milieu requires a joint analysis of how science, technology and the global market operate, which may provide a new interpretation of environmental issues, since changes in Nature are subordinated to those logics (Santos 1996: 159). For Santos, information is the power that drives these forces. A global logic transmits stored and coded information in actions and objects 'dominating all territories and each territory as a whole' (Santos 1996: 160). Until two centuries ago, every society used and managed a section of Nature, but the latest technologic leap and, more recently, the globalization of the economy, 'has made possible the unification of Nature, by adopting a unique technical model' (Santos 1995: 697), which serves a global economic logic. Consequently, exploitation of resources and environmental damages are also unified with space. This is one of the greatest contributions of Santos: he relates a global spatial logic to global environmental crisis, which is not usually included in official environmentalist discourse. Interrogating spatial logics and their effects on technical environments reveals the real causes of the current environmental upheavals.

12.5   Ethical

It is a big paradox—and a tragedy—that the increasing worldwide environmental concern has not helped to rectify many of our unsustainable behaviours. It would appear contradictory to think that widespread acknowledgement of environmental issues is, in the end, responsible for maintaining and, in fact, worsening environmental illness. Unfortunately, there is some truth in this and Santos, once more, may shed some light on this controversy.

For Santos, a particular image, or aesthetics, of Nature is projected by mass media and shapes attitudes toward environmental problems. The mass media is

---

[9]Wright (2004) refers to 'progress trap' as an innate condition of the historical evolution of societies: in pursuing progress through human ingenuity, human societies introduce problems they do not have the resources or political will to solve. Kingsnorth (2013: 53) explains it as follows: 'each improvement in our knowledge or in our technology will create new problems, which require new improvements. Each of these improvements tends to make society bigger, more complex, less human-scale, more destructive of nonhuman life, and more likely to collapse under its own weight'.

central to Santos because it disperses information that is assimilated by large and diverse sectors of society. As he describes, 'perception is mutilated whenever the media consider it necessary; attracting attention through fear and sensationalism. Many environmentalist movements, driven by media, destroy, mutilate, and repress Nature' (Santos 1992: 102). The image of Nature usually broadcast in the mass media has two aesthetics: a positive aesthetics, exalting and stressing the wilderness or what Santos calls 'fantasy'; and a negative aesthetics, presenting a desolated, polluted and dark face of Nature, or what he calls 'fear' (Santos 1992).

Asserting the importance of temporal and spatial contextualization, Santos argues that 'spectacle nature substitutes historic nature' (Santos 1992: 102). In other words, spectacle nature displaces that nature which is livelihood, 'the place of labour of all humans' (Santos 1992: 102). It is usual in the media but also in political discourse to find a yearning for a pristine Nature that humans have never known; they offer a sense of affection for a stereotyped idea of Nature, mobilizing images of beautiful landscapes, friendly wildlife, and healthy and pure environments. According to Santos's ontology, these technical objects, that is, the mass media milieu, also have intentions, intrinsic values and moral values, which associate the good, the correct and, thus, the desirable with such aesthetic Nature. However, as Santos denounces: 'The power of images tends, therefore, to wipe out the concepts, empty them of their respective meanings' (Santos 1995: 702).

Climate change provides us with perhaps the best example of the way the media use images to dramatically caricature Nature. For instance, the award-winning Al Gore documentary *An Inconvenient Truth* (2006) presents pairs of pictures of the same place from different years in order to convince the audience that glaciers in different regions are receding. Wide shots of landscape degradation transmit veracity. Impressive scenes of flooding, fire and evacuation are shown in order to provoke strong emotions, whether the images are directly related to climate change or not.

Following Santos, technical objects such as these films have specific intentions, which denaturalize and fragment the complexity of Nature itself. In short, the denaturalization of Nature is the dehumanization of Nature, insofar as human dimensions and human environments are not given the same prominence and equivalent treatment as wilderness: 'media…threatens the integrity of people…and precisely, the human factor must be recovered by an environmentalism whose main motivation should not be emotion' (Santos 1995: 703). Accordingly, nature and environmental problems are part of human affairs, insofar as humanity interacts inevitably and constantly with the global biophysical system, that is, the biosphere. Humans need specific environmental conditions to survive and natural resources to produce technical objects, which allow sustenance and societal progress. This is Santos's premise for a proper ethics with respect to the environment in the transition to more sustainable societies.

12.6 Final Remarks: Santos and the Environmental Crisis as 'Deterritorialization'

As I mentioned in the introduction, environmental issues were not one of Milton Santos's central concerns, save for his two aforementioned essays. However, his broad and holistic theoretical approach encompasses the main ontological dimensions that intermediate between humans and Nature. He presents coherent views of human environment or *milieu*, where space operates as a crucial dimension. He offers tools for a distinctive analysis of the environmental crisis, being specifically and originally geographic. In addition, Santos's thought is versatile and cross-disciplinary, as well as both critical and radical.

In line with his thinking, a proper framework for the study of the environmental crisis must include a broader sense of environment as *milieu* or geographic environment. Santos believes that 'the problem of human space has today a magnitude it never had before' (Santos 1992: 97), and that understanding the association between problems of the environment and problems of human space is imperative. Hegemonic agents of globalization operate within a model of development that has severely affected our way of conceiving and handling Nature, a model that operates within the new technical-scientific-informational milieu. A globalized economic logic is often indifferent to 'local and environmental realities,' which 'explains why the environmental crisis has taken place in the current historical period, because the power of triggered forces in a place overcomes the local capacity to control them' (Santos 1996: 170). Environmental damages are thus locally produced, but transported by techniques moved by distant interests (Santos 1996: 170) in what Santos (1996) calls deterritorialization. A globalized economic logic is behind our sense of instability and the constant movement of people, capital and information. Our links with objects, people and places are increasingly ephemeral and unstable (Bauman 2000). In turn, the global environmental crisis may be depicted through the metaphor of 'ecological uprooting' (Toro 2011: 645). People, traditional uses, cultural artifacts and activities operating in balance with the physical limitations of local environments are being replaced by delocalized and expansive patterns. The ecological disaster, according to Santos, is thus deterritorialized and even globalized.

Santos's interpretation of environmental crisis is *real*, that is, situational, contextual, non-apocalyptical and non-speculative. In addition, it is ontologically geographically-based, emphasizing those features which constitute an artificial- *and* natural-made milieu. Focusing on human environment or space requires moving beyond disciplinary biases or geographic corporatisms, and being guided by veracity: 'efforts [that] concern us…are typical of an age which illustrates risks… and the need for, in all fields of knowledge, acting heroically, while pursuing the truth' (Santos 1992: 105).

References

Amin, S. (1997). *Capitalism in the age of globalization*. London: Zed Books.

Bauman, Z. (2000). *Liquid modernity*. Cambridge: Polity.

Beck, U., & Willms, J. (2003). *Conversations with Ulrich Beck*. Cambridge: Polity Press.

Blackmore, J. (2001). Universities in crisis? Knowledge economies, emancipator pedagogies, and the critical intellectual. *Educational Theory, 51*(3), 353–370.

Boghossian, P. (2006). *Fear of knowledge: Against relativism and constructivism*. New York: Oxford University Press.

Buttimer, A. (1993). *Geography and the human spirit*. Baltimore-London: The Johns Hopkins University Press.

Capra, F. (1982). *The turning point: Science, society, and the rising culture*. New York: Bantam Books.

Crutzen, P. J., & Stoermer, E. F. (2000). The 'Anthropocene'. *Global Change Newsletter, 41*, 17–18.

Escobar, A. (2015). Degrowth, postdevelopment, and transitions: A preliminary conversation. *Sustainability Science, 10*, 451–462.

Feyerabend, P. (1975). *Against method*. London: NBL.

Funtowicz, S. O., & Ravetz, J. R. (1993). Science for the post-normal age. *Futures, 25*(7), 739–755.

Galcerán, M. (2010). La educación universitaria en el centro del conflicto. In Edu-Factory & Universidad Nómada (Eds.), *La Universidad en conflicto. Capturas y fugas en el mercado global del saber* (pp. 13–39). Madrid: Traficantes de Sueños.

Goudie, A. (2006). *The human impact on the natural environment. Past, present, and future*. 6th ed, Malden-Oxford-Victoria: Blackwell Publishing.

Hernández del Águila, R. (1985). *La crisis ecológica: ¿de dónde viene, a dónde nos conduce?* Barcelona: Laia.

Kingsnorth, P. (2013). Dark ecology: Searching for truth in a post-green world. *Orion Magazine*, Jan–Feb. Retrieved August 30, 2016, from http://www.orionmagazine.org/index.php/articles/article/7277.

Kuhn, T. S. (1962). *The Structure of Scientific Revolutions*. Chicago: University of Chicago Press.

Krimsky, S. (2003). *Science and the private interest. Has the lure of profits corrupted biomedical research?* Oxford: Rowman & Littlefield.

Lakatos, I. (1978). *The methodology of scientific research programmes: Philosophical papers volume 1*. Cambridge: Cambridge University Press.

Lander, E. (2008). La ciencia neoliberal. *Tabula Rasa, 9*, 247–283.

Latour, B. (1991). *Nous n'avons jamais été modernes; Essai d'anthropologie symétrique*. Paris: La Découverte: 'L'armillaire'.

Leff, E. (Ed.) (1986). *Los problemas del conocimiento y la perspectiva ambiental del desarrollo*. México: Siglo XXI Editores.

Levitt, T. (1983). The globalization of markets. *Harvard Business Review, 61*(3), 92–102.

Martínez, M. (1993). *El paradigma emergente: Hacia una nueva teoría de la racionalidad científica*. Barcelona: Gedisa Editorial.

Melgaço, L. (2013). Security and surveillance in times of globalization: An appraisal of Milton Santos' theory. *International Journal of E-Planning Research, 2*(4), 1–12.

Moncrief, L. W. (1970). The cultural basis for our environmental crisis. *Science, 170*, 508–512.

Morin, E. (1977[2004/2008]). *La Méthode (6 volumes)*. Seuil Opus: Paris.

Morin, E. (1999). *Seven complex lessons in education for the future*. Paris: United Nations Educational, Scientific and Cultural Organization.

Naredo, J. M. (2006). *Raíces económicas del deterioro ecológico y social: Más allá de los dogmas*. Madrid: Siglo XXI.

Norgaard, R. B. (1994). *Development betrayed: The end of progress and coevolutionary revisioning of the future*. London: Routledge.

Peluso, M. L. (2013). The challenge of understanding nature in Milton Santos' Work. *Revista Eletrônica: Tempo-Técnica-Territorio*, 4(1), 20–27.

Popper, K. (1965). Normal science and its dangers. In I. Lakatos & A. Musgrave (Eds.), *Criticism and the growth of knowledge. Proceedings of the International Colloquium in the Philosophy of Science, London, 1965* (Vol. 4, pp. 51–58). Cambridge: Cambridge University Press.

Porritt, J. (2000). *Playing safe: Science and the environment*. London: Thames & Hudson.

Redclift, M. (1984). *Development and the environmental crisis: Red or green alternatives*. London; New York: Methuen.

Robertson, R. (1995). Glocalization: Time-space and homogeneity-heterogeneity. In M. Featherstone, S. Lash & R. Robertson (Eds.), *Global modernities*. London, Sage.

Robertson, R., & Lechner, F. (1985). Modernization, globalization and the problem of culture in World-Systems theory. *Theory, Culture & Society*, 2(3) 103–117.

Sachs, W. (2010). Globalization, convergence and the Euro-Atlantic development model. In M. Redclift & G. Woodgate (Eds.), *The international handbook of environmental sociology* (pp. 262–275). Cheltenham, UK: Edward Elgar Published Limited.

Sack, R. D. (1997). *Homo geographicus: A framework for action, awareness, and moral concern*. Baltimore: The Johns Hopkins University Press.

Santos, M. (1985). *Espaço e Método*. São Paulo: Nobel.

Santos, M. (1992). A redescoberta da natureza. *Estudos Avançados*, 6(14), 95–106.

Santos, M. (1995). A questão do meio ambiente: desafios para a construção de uma perspectiva transdisciplinar. *Anales de Geografía de la Universidad Complutense*, 15, 695–706.

Santos, M. (1996). *A Natureza do espaço: Técnica e tempo; Razão e emoção*. São Paulo: Hucitec.

Schumacher, E. F. (1973). *Small is beautiful: A study of economics as if people mattered*. London: Blond & Briggs.

Séris, J.-P. (1994). *La technique*. Paris: Presses Universitaires de France.

Toro, F. (2011). *Crisis ecológica y geografía. planteamientos y propuestas en torno al paradigma ecológico-ambiental*. Granada: Editorial Universidad de Granada.

Tuan, Y. F. (1974). *Topophilia: A study of environmental perception, attitudes, and values*. Englewood Cliffs: Prentice-Hall.

Turner, B. L., et al. (1990). Two types of global environmental change: Definitional and spatial-scale issues in their human dimensions. *Global Environmental Change*, 1, 14–22.

Wagensberg, J. (2014). On the existence and uniqueness of the scientific method. *Biological Theory*, 9(3), 331–346.

White, L. (1967). The historical roots of our ecological crisis. *Science*, 155, 1203–1207.

Wright, R. (2004). *A short history of progress*. Edinburgh: Canongate Books.

Milton Santos's Selected Bibliography

This is a selection of some of Milton Santos's most relevant works organized by type and language. For a more complete list of Santos publications see Santos (2001).

Books (In Portuguese)

Santos, M. (1948). *O Povoamento da Bahia: Suas causas econômicas.* Salvador: Imprensa Oficial da Bahia.

Santos, M. (1959). *O centro da cidade do salvador.* Salvador: Livraria Progresso Editora.

Santos, M. (1960). *Marianne em preto e branco.* Salvador: Livraria Progresso Editora.

Santos, M. (1965). *A Cidade nos países subdesenvolvidos.* Rio de Janeiro: Civilização Brasileira.

Santos, M. (1978). *Por uma geografia nova.* São Paulo: Hucitec.

Santos, M. (1978). *O trabalho do geógrafo no terceiro mundo.* São Paulo: Hucitec.

Santos, M. (1978). *A pobreza urbana.* São Paulo: Hucitec.

Santos, M. (1978). *O espaço dividido.* Rio de Janeiro: Livraria Editora Francisco Alves.

Santos, M. (1978). *Economia espacial: Críticas e alternativas.* São Paulo: Hucitec.

Santos, M. (1979). *Espaço e sociedade.* Petrópolis: Vozes.

Santos, M. (1980). *A urbanização desigual.* Petrópolis: Vozes.

Santos, M. (1981). *Manual de geografia urbana.* São Paulo: Hucitec.

Santos, M. (1982). *Pensando o espaço do homem.* São Paulo: Hucitec.

Santos, M. (1982). *Ensaios sobre a urbanização latino-americana.* São Paulo: Hucitec.

Santos, M. (1985). *Espaço e método.* Sao Paulo: Nobel.

Santos, M. (1987). *O Espaço do cidadão.* Sao Paulo: Nobel.

Santos, M. (1988). *Metamorfoses do espaço habitado.* São Paulo: Hucitec.

Santos, M. (1990). Metrópole corporativa fragmentada: O caso de São Paulo. Sao Paulo: Nobel.

Santos, M. (1993). *A urbanização brasileira.* São Paulo: Hucitec.

© Springer International Publishing AG 2017
L. Melgaço and C. Prouse (eds.), *Milton Santos: A Pioneer in Critical Geography from the Global South*, Pioneers in Arts, Humanities, Science, Engineering, Practice 11, DOI 10.1007/978-3-319-53826-6

Santos, M. (1994). *Por uma economia política da cidade*. São Paulo: Hucitec - Editora PUC-SP.

Santos, M. (1994). *Técnica, espaço, tempo: Globalização e meio técnico-científico informacional*. São Paulo: Hucitec.

Santos, M. (1996). *A Natureza do espaço: Técnica e tempo; Razão e emoção*. São Paulo: Hucitec.

Santos, M. (2000). *Território e sociedade*: Entrevista com Milton Santos. São Paulo: Fundação Perseu Abramo.

Santos, M. (2000). *Por uma outra globalização: Do pensamento único à consciência universal*. Rio de Janeiro: Record.

Santos, M., & Silveira, M. L. (2001). *O Brasil: Território e sociedade no início do século XXI*. Rio de Janeiro, Brazil: Record.

Books (In French)

Santos, M. (1970). *Dix essais sur les villes des pays sous-développés*. Paris: Ophrys.

Santos, M. (1971). *Le métier du géographe en pays sous-développés*. Paris: Ophrys.

Santos, M. (1971). *Les villes du tiers monde*. Paris: M.-Th. Génin, Librairies Techniques.

Santos, M. (1975). *L'espace partagé*. Paris: M.-Th. Génin, Librairies Techniques.

Santos, M. (1985). *Pour une géographie nouvelle*. Paris: Editions Publisud.

Santos, M. (1990). *Espace et méthode*. Paris: Publisud.

Santos, M. (1997). *La nature de l'espace: Technique et temp; Raison et émotion*. Paris: L'Harmattan.

Books (In Spanish)

Santos, M. (1973). *Geografia y economia urbanas en los países subdesarrolados*. Barcelona: Oikos-Tau.

Santos, M. (1986). *Espacio y metodo*. Barcelona: Geocritica, 65, Septiembre, Universidad de Barcelona.

Santos, M. (1990). *Por una geografia nueva*. Madrid: Espasa-Calpe.

Santos, M. (1996). *De la totalidad al lugar*. Barcelona: Oikos-Tau.

Santos, M. (2000). *La naturaleza del espacio*: Técnica y tiempo; Razón y emoción. Barcelona: Ariel.

Santos, M. (2004). *Por otra globalización: Del pensamiento único a la conciencia universal*. Bogotá: Convenio Andrés Bello.

Book (In English)

Santos, M. (1979). *The shared space: The two circuits of the urban economy in underdeveloped countries*. Translated by C. Gerry. London and New York, Methuen.

Santos, M. (2017). *Toward an other globalization: From the single thought to universal conscience*. Translated by L. Melgaço and T. Clarke. Cham: Springer.

Book Chapters (In Portuguese)

Santos, M. (1960). Geografia e desenvolvimento econômico. In M. Neto & A. Luís (Eds.), *Desenvolvimento: Problemas e soluções* (pp. 107–126). Salvador: Imprensa Oficial de Bahia.

Santos, M. (1982). Geografia, marxismo e subdesenvolvimento. In R. Moreira (Ed.), *Geografia: teoria e crítica* (pp. 13–22). Petropolis: Vozes,

Santos, M. (1991). Involução metropolitana e economia segmentada. In: A. C. T. Ribeiro & D. B. Pinheiro Machado (Eds.), *Metropolização e rede urbana: Perspectivas dos anos 90*. Rio de Janeiro: IPPUR, UFRJ.

Santos, M. (1994). O Retorno do território. In M. Santos, M. A. de Souza & M. L. Silveira (Eds.), *Território: Globalização e fragmentação* (pp. 15–20). São Paulo: Hucitec and ANPUR.

Santos, M. (1996). As Cidadanias mutiladas. In J. Lerner (Ed.), *O Preconceito* (pp. 133–144). São Paulo: Imprensa Oficial do Estado.

Santos, M. (2000). A era da inteligência baseada na máquina. In A. L. Trindade & R. Santos (Eds.), *Multiculturalismo - mil e uma faces da Escola* (pp. 149–157), Rio de Janeiro: DP&A.

Santos, M. (2000). Globalização e meio geográfico: Do mundo ao lugar. In A. J. Souza, E. Clemente de Souza & L. Magnomi Jr. (Eds.), *Paisagem, território, região: Em busca da identidade* (pp. 51–56). Cascavel: Edunioeste.

Book Chapters (In English)

Santos, M. (1973). Economic development and urbanization in underdeveloped countries: The two-flow systems of the urban economy and their spatial implications. In D. McKee & W. Leahy (Eds.), *Urbanization and the development process*. New York: The Free Press.

Santos, M. (1979). Circuits of work. In S. Wallman (Ed.), *Ethnicity at work* (pp. 215–226). London: The Macmillan Press.

Santos, M. (1995). World time and world space or just hegemonic time and space? *Geography, History and Social Sciences* (pp. 45–49). Dordrecht: Kluwer Academic Publishers.

Santos, M. (1995). Contemporary acceleration: World-time and world-space. In G. Benko & U. Strohmayer (Eds.), *Geography, History and Social Sciences* (pp. 171–176). Dordretch: Kluwer Academic Publishers.

Santos, M. (1996). São Paulo: A growth process full of contradictions. In A. Gilbert (Ed.), *The mega-city in Latin America*. Tokyo, New York, Paris: United Nations University Press.

Articles (In Portuguese)

Santos, M. (1982). O Espaço e seus elementos: Questões de método. *Revista Geografia e Ensino. 1*, 19–30.

Santos, M. (1982). Para que a geografia mude sem ficar a mesma coisa. *Boletim Paulista de Geografia, 59*, 5–22.

Santos, M. (1985). O período técnico-científico e os estudos geográficos. *Revista do Departamento de Geografia da USP, 4,* 15–20.

Santos, M. (1989). O espaço geográfico como categoria filosófica. *Terra Livre, 5,* 9–20.

Santos, M. (1991). A revolução tecnológica e o território: Realidades e perspectivas. *Caderno Prudentino de Geografia,* 13.

Santos, M. (1991). Flexibilidade tropical. *Arquitetura e Urbanismo, 38,* 44–45.

Santos, M. (1992). 1992: A redescoberta da Natureza. *Estudos Avançados, 14*(6), 95–106.

Santos, M. (1993). Metrópole: A força dos fracos é o seu tempo lento. *Ciência e Ambiente, 5*(7), 7–12.

Santos, M. (1993). Objetos e ações: Dinâmica espacial e dinâmica social. *Geosul, 14,* 49–59.

Santos, M. (1996). Por uma geografia cidadã: Por uma epistemologia da existência. *Boletim Gaúcho de Geografia. 21,* 7–14.

Santos, M. (1996). O Lugar: encontrando o futuro. *RUA-Revista de Arquitetura e Urbanismo, 6,* 34–39.

Santos, M. (1997). Da política dos estados à política das empresas. *Cadernos da Escola do Legislativo, 3*(6), 9–23.

Santos, M. (1999). O Território e o saber local: Algumas categorias de análise. *Cadernos IPPUR, 3*(2), 15–26.

Santos, M. (1999). O dinheiro e o território. GEOgraphia, *1*(1), 7–13.

Santos, M., et al. (2000). O Papel ativo da geografia: Um manifesto. *Território, 5* (9), 103–109.

Articles (In English)

Santos, M. (1974). Geography, marxism and underdevelopment. *Antipode, 6*(3), 1–9.

Santos, M. (1975). The periphery at the pole: Lima, Péru. In G. Grappert & H. N. Rose (Eds.), *The Social Economy of Cities, Urban Affairs Annual Review, 9,* 335–360.

Santos, M. (1975). Space and domination: a Marxist Approach. *International Social Science Journal, 27*(2), 346–363.

Santos, M. (1975). Underdevelopment, growth pole and social justice. *Civilisations, 25*(1–20), 18–31.

Santos, M. (1976). Articulation of modes of production and the two circuits of urban economy wholesalers in Lima, Peru. *Pacific Viewpoint, 3,* 23–36.

Santos, M. (1976). Economic development and urbanization in underdeveloped countries: The two sub-systems of the urban economy. *Journal of the Geographical Association of Tanzania,* 6–36.

Santos, M. (1977). Society and space: Social formation as theory and method. *Antipode, 9*(1), 3–13.

Santos, M. (1977). Spatial dialectics: The two circuits of urban economy in underdeveloped countries. *Antipode, 9*(3), 49–60.

Santos, M. (1977). Planning underdevelopment. *Antipode*, *9*(3), 86–98.

Santos, M. (1980). The devil's totality. *Antipode*, *12*(3), 41–46.

Santos, M. (1995). Universal reason, local reason: The spaces of rationality. *GeoJournal*, *36*(1), 108–110.

Santos, M. et al. (2017). The active role of geography: A manifesto. Translated by T. Clarke & L. Melgaço. *Antipode*, *49*(4). Retrieved February 1, 2017, from http://onlinelibrary.wiley.com/doi/10.1111/anti.12318/abstract.

Articles in Spanish

Santos, M. (1984). La Geografia a fines del siglo XX: Nuevas funciones de una disciplina amenazada. *Revista Internacional da Ciencia Sociales*, *102*(36), 693–709.

Santos, M. (1991). Modernisación, medio tecnico científico y urbanización en Brasil. *Anales de Geografia de la Universidad Complutense*, *10*, Madri.

Santos, M. (1993). El mundo y la geografía hoy. *Revista Geográfica Venezolana*, *33*, (1), 5–9.

Santos, M. (1995). Los espacios de la globalización. *Revista Universidad del Valle*, *10*, 36–41.

Santos, M. (1997). Nuevas concepciones de la Geografia. *GeoUruguay*, *1*(set), 117–123.

Santos, M. (1997). La fuerza del lugar: Orden universal, orden local *Geographikós*. *8*(7).

Articles in French

Santos, M. (1959). Quelques problemes geographiques du centre de la ville de salvador. *L'Information Géographique*, *23*, 93–98.

Santos, M. (1968). La géographie urbaine et l'économie des villes dans les pays sous-développés. *Revue de Géographie de Lyon*, *4*(43), 361–376.

Santos, M. (1969). De la géographie de la faim à la planification régionale. *Revue Tiers Monde*, *37*, 95–114.

Santos, M. (1972). Les villes incomplètes des pays sous-développés. *Annales de Géographie*, *445*, 316–323.

Santos, M. (1974). Sous-dévéloppment et poles de croissance économique et sociale. *Revue Tiers Monde*, *58*, 271–286.

Santos, M. (1977). Une géographie de la médecine. *Hérodote* – Des résponses aux questions de Michel Foucault, *6*, 28–29.

Santos, M. (1978). De la société au paysage: La signification de l'espace humain, *Hérodote*, *9*, 66–73.

Santos, M. (1980). Société et espace transnationalisés dans le Venezuela actuel. *Revue Tiers Monde*, *21*(84), 709–720.

Santos, M. (1981). Structure, totalité, temps: L'espace du monde d'aujourd'hui. *Espaces-Temps*, *18–19–20*, 103–122.

Santos, M. (1988). Reflexions sur le rôle de la géographie dans la periode technico scientifique. *Cahiers de Géographie du Quebec*, *32*(87), 313–319.

Santos, M. (1992). Temps-monde et espace-monde: Relever le défi conceptual. *Strates, 7*.

Santos, M. (1993). Les espaces de la globalisation. *Cahiers du GEMDEV, 20*, 161–172.

Full Texts in Proceedings of Conferences (in English)

Santos, M. (1970). The urbanisation of underdeveloped countries and its effects on the nutrition of urban populations. *Proceedings of the 3rd International Congress on Food, Science and Technology*, Washington D.C., Aug (pp. 169–174).

Santos, M. (1973). Urban crisis or epiphenomenon? *Proceedings of the International Population Conference*, Liège, Belgium (pp. 287–291).

Santos, M. (1979). The cities of the third world: Industrialization and spatial repercussions. Proceeding of the 22nd International Geographical Congress, Montreal, Canada (pp. 10–17).

Santos, M. (1979). Research for the urban future: The case of Latin America. *Proceedings of the 22nd International Geographical Congress*, Ottawa, Canada (pp. 125–129).

Santos, M. (1992). Management and planning of great metropolis of the Third World. *Acts of the Second Conference of the World Capitals*, Dakar (pp. 44–48).

Santos, M. (1992). Issues concerning the capitals of the world: capital cities in developing countries, Foreword. *Acts of the Second Conference of the World Capitals*, Dakar (pp. 44–48).

Other Publication in English

Santos, M. (1975). Underdevelopment and poverty: A geographer's view. Toronto: The Latin American in Residence Lectures 1972–1973, University of Toronto.

About Santos (In Portuguese)

Costa, P. E. F. (2013). O "jovem Milton Santos": personagem do protótipo metodológico: Revelar [matrizes clássicas originárias] para definir [vanguarda, universalidade e viés geográfico]. (Ph.D. dissertation). Rio Claro: UNESP.

Ferreira, A. M. de A. (2000). *Para um vocabulário fundamental da obra de Milton Santos: Com equivalência em francês*. (Ph.D. dissertation). São Paulo: University of São Paulo.

Grimm, F. (2011). *Trajetória epistemológica de milton santos*. (Ph.D. dissertation). São Paulo: University of São Paulo. Retrieved August 25, 2016, from http://www.teses.usp.br/teses/disponiveis/8/8136/tde-26062012-143800/pt-br.php.

Moraes, A. C. R. (2013). *Território na geografia de Milton Santos*. São Paulo: Annablume.

Silva, M. A. (2002). Milton Santos: a trajetória de um mestre. In *El Ciudadano, La Globalización y la Geografía. Homenaje a Milton Santos. Scripta Nova*, 2(124). Retrieved August 18, 2016, from http://www.ub.edu/geocrit/sn/sn-124f.htm.

Silva, F. S., & Silva, M. A. (2004). Uma leitura de Milton Santos (1948–1964). *Geosul, 19*(37), 157–189.

Souza, M. A. (Ed.) (1996). *O Mundo do cidadão: Um cidadão do mundo*. São Paulo: Hucitec.

Tendler, S. (2006). *Encontro com Milton Santos ou o mundo global visto do lado de cá*. Rio de Janeiro: Caliban Produções, DVD (89 min), son., color.

Tiercelin dos Santos, M.-H., & Levy, J. (2011). Biografia. *Milton Santos* [website]. Retrieved August 18, 2016, from http://miltonsantos.com.br/site/biografia/.

Vasconcelos, P. A. (2001). Milton Santos, geógrafo e cidadão do mundo (1926–2001), *Afro-Ásia, 25–26*, 369–405.

About Santos (In English)

Melgaço, L. (2013). Security and surveillance in times of globalization: An appraisal of Milton Santos' Theory. *International Journal of E-Planning Research, 2*(4), 1–12.

Melgaço, L. (2017). Thinking Outside the Bubble of the Global North: Introducing Milton Santos and "The Active Role of Geography". *Antipode, 49*(4), Retrieve 5 March 2017, from http://onlinelibrary.wiley.com/doi/10.1111/anti.12319/full.

Melgaço, L & Clarke, T. (2017). Introducing Milton Santos and "The Active Role of Geography". Symposium. Antipode. Retrieved 5 March 2017, from https://antipodefoundation.org/supplementary-material/the-active-role-of-geography/.

Souza, M. A. (2009). Santos, M. In R. Kitchin & N. Thrift (Eds.), *International Encyclopaedia of Human Geography* (pp. 11–14). Oxford: Elsevier.

About Santos (In Other Languages)

Levy, J. (Ed.) (2007). *Milton Santos, philosophe du mondial, citoyen du local*. Paris: Presses Polytechniques et Universitaires Romandes.

Maurel, J. (1996). Homenaje al profesor Milton Santos. *Anales de Geografía de la Universidad Complutense, 16*, 203–223.

Scripta N. (2002). El ciudadano, la globalización y la geografía: Homenaje a Milton Santos, 6 (124). Retrieved August 18, 2016, from http://www.ub.edu/geocrit/sn/sn-124.htm.

Reference

Santos, M. (2001). Curriculum Vitae. *Milton Santos* [website]. Retrieved August 18, 2016, from http://www.miltonsantos.com.br/site/miltonsantos_curriculum.pdf.

About Milton Santos

Milton Almeida dos Santos (1926–2001) was one of the most important and prominent geographers in the so-called Global South. Despite his specialization in themes such as urban studies, developing countries and globalization, his main contributions to human science are his expansive theorizations operating well beyond any specific subdiscipline of geography, creating what could be called today a 'Miltonian School of Geographic Thought'. He assembled a set of coherent and complementary concepts and ideas that together explain different spatial phenomena.

Born in Brotas de Macaúbas, Brazil, Santos first obtained a degree in law, but his passion from his earliest professional years was geography. He worked in high schools as a teacher of geography, in journalism and in politics before starting his academic life as a geographer. He completed his Ph.D. in Geography in 1958 at the University of Strasbourg, France, under the supervision of Prof. Jean Tricart. Returning to Brazil, Santos worked in different universities until 1964, when a military coup d'état ousted Brazil's democratically elected government. Santos was arrested by the military police and was released on the condition that he be deported. During 13 years of exile Santos worked as an academic in different countries in Europe, North America and Africa.

In 1977 Santos returned again to Brazil and from 1983 onwards was affiliated with the University of São Paulo where he worked until he passed away on 24 June 2001 at the age of 75. Santos received the *honoris causa* distinction from 20 universities. His impressive academic production includes almost 300 scientific articles and more than 40 books, including classics like *The Shared Space, A Natureza do Espaço* and *Toward an Other Globalization*. Milton Santos was the only geographer from outside the Euro-Anglo world to be awarded the Vautrin Lud International Prize, considered the Nobel Prize of Geography.

© Springer International Publishing AG 2017
L. Melgaço and C. Prouse (eds.), *Milton Santos: A Pioneer in Critical Geography from the Global South*, Pioneers in Arts, Humanities, Science, Engineering, Practice 11, DOI 10.1007/978-3-319-53826-6

About the Editors

Lucas Melgaço is an Assistant Professor in the Department of Criminology of the Vrije Universiteit Brussel (VUB), Belgium where he combines his background in geography with his specialization in security, surveillance and policing studies. He holds a doctorate degree in Geography from a partnership between the University of São Paulo (USP) and the University of Paris 1—Panthéon Sorbonne with a thesis entitled 'Urban securitization: from the psychosphere of fear to the technosphere of security'. Lucas is a former post-doctoral researcher at the Crime and Society Research Group (CRiS) at the VUB, at the Surveillance Studies Centre at Queen's University, Canada and at the Department of Geography of the Federal University of Rio de Janeiro (UFRJ). His main scientific interests are in the domains of urban security, urban conflicts, surveillance, public order, social movements and protests, and in the relationships between information and communication technologies and security. He has also worked on translating and introducing the theories of Milton Santos to the English-speaking community. Lucas is editor-in-chief of the journal *Criminological Encounters*.

Address: Lucas Melgaço, Pleinlaan 2, 1050 Brussels, Belgium
Email: lucas.melgaco@vub.ac.be

Carolyn Prouse is a doctoral candidate in the Department of Geography at the University of British Columbia (UBC) in Vancouver, Canada. She holds a Master's degree and bachelor's degree from Queen's University in Kingston, Canada. Carolyn has spent time at the University of São Paulo and at the cultural and research institute Barraco #55 in Complexo do Alemão, Rio de Janeiro, Brazil. Her research interests are in urban studies, critical development studies and critical health geographies. She operates conceptually at the nexus of postcolonial/decolonial and anti-racist feminist approaches.

Address: Carolyn Prouse, 210D-1984 West Mall, Vancouver, BC V6T 1Z2, Canada
Email: carolyn.prouse@geog.ubc.ca

About the Contributors

Sarita Albagli is a sociologist with a Ph.D. in Geography. She is a Senior Researcher at the Brazilian Institute of Information in Science and Technology (IBICT), and Senior Lecturer in the Post-Graduate Program in Information Science, developed by IBICT and the Federal University of Rio de Janeiro (UFRJ). She coordinates the Interdisciplinary Laboratory on Information and Knowledge Studies (Liinc) and she is Editor of the scientific journal *Liinc em Revista*. Her research interests include the following topics: information, knowledge and social innovation; open and collaborative science and technology; network and territorial informational dynamics.
Email: sarita@ibict.br

Aurélien Reys is a geographer who specializes in Development Geography. He wrote his Ph.D. dissertation on the Brazilian gemstones industry. He is currently a Post-doctoral Researcher at the Centre de Coopération International en Recherche Agronomique pour le Développement (CIRAD). His current research focuses on the impacts of agricultural investment on household livelihoods and food security in East Africa.
Email: aurelien.reys@cirad.fr

Eliza Pinto de Almeida holds bachelor, Master's and Ph.D. degrees in Geography from the University of São Paulo (USP). She is Associate Professor in the Institute of Geography, Development and Environment at the Federal University of Alagoas. Her scientific interests include the relationships between territory and health with an emphasis on the organization of public health services.
Email: eliza.almeida@igdema.ufal.br

Samuel Frederico has graduate (2002) and Master's (2004) degrees in Geography from the State University of Campinas, and a Ph.D. (2009) in Human Geography from the University of São Paulo, with a doctoral internship (2007/08) held at the Université de Toulouse II Le Mirail, France. He is currently Assistant Professor at the São Paulo State University (UNESP—Rio Claro) and Assistant Professor of the Graduate Program in Geography at the State University of Campinas (Unicamp). Samuel's research is in Economic and Regional Geography, and focuses on the following themes: modernization and expansion of Brazilian agriculture;

© Springer International Publishing AG 2017
L. Melgaço and C. Prouse (eds.), *Milton Santos: A Pioneer in Critical Geography from the Global South*, Pioneers in Arts, Humanities, Science, Engineering, Practice 11, DOI 10.1007/978-3-319-53826-6

competitive regions; urban-rural relationships; agro-logistics; the sugarcane industry; coffee production; soybeans; and financial capital and agribusiness.
Email: sfrederico@rc.unesp.br

Marina Castro de Almeida has a graduate (2002) degree in Geography from the Pontifical Catholic University of São Paulo, a Master's (2005) in Geography from the State University of Campinas, and Ph.D. (2013) in Human Geography from the University of São Paulo, with a doctoral internship (2012) held at the City University of New York (CUNY). She is currently Assistant Professor at the Minas Gerais Triangle Federal University (UFTM). Her research in Geography has an emphasis in Economic Geography, focusing on the following themes: globalization; uneven geographical development; territorial distribution of business services; and precarious work.
Email: marinacastrodealmeida@gmail.com

Luís Angelo dos S. Aracri holds bachelor, Master's and doctoral degrees in Geography from the Federal University of Rio de Janeiro (UFRJ). Since 2011 he has been Associate Professor at the Federal University of Juiz de Fora (UFJF). His scientific interests include: technological transformations and their impacts on the organization of territory; dynamics of regional productive structures; and the political economy of the territory.
Email: luis.aracri@ufjf.edu.br

Fabrício Gallo holds bachelor, Master's and doctoral degrees in Geography from the State University of Campinas (UNICAMP), Brazil. He is an Assistant Professor in the Department of Territorial Planning and Geoprocessing (DEPLAN) of Universidade Estadual Paulista (UNESP). His scientific interests include Brazilian territorial integration with emphases on federalism; intergovernmental transfers of public resources; and territorial politics.
Email: fgallo@rc.unesp.br

Júlia Adão Bernardes is a Professor in the Department of Geography at the Federal University of Rio de Janeiro (UFRJ), where she also completed her bachelor (1974), and Master's (1983) degrees. She holds a Ph.D. (1993) in Human Geography from the Universitat de Barcelona, Spain. Julia's interests include economic geography; regional geography; and modernization of Brazilian agriculture.
Email: julia.rlk@gmail.com

Roberta Carvalho Arruzzo holds bachelor (2002), Master's (2004) and doctoral (2009) degrees in Geography from the Federal University of Rio de Janeiro (UFRJ). In 2011 and 2012 she engaged in teaching and research activities at the same university. She is currently Assistant Professor at the Rural Federal University of Rio de Janeiro, Coordinator of the Research Group of Territorial Resistance (COLETIVO), and a researcher with the research group Geography and Indigenous Peoples (Geopovos) at the same institution. She has experience in the fields of Human Geography and Agricultural Geography and for more than ten years has

been studying the territorial relations between the modernization of agriculture and indigenous peoples.
Email: roberta.arruzzo@pq.cnpq.br

Maria Tereza Duarte Paes graduated (1985) from Geography at the Universidade Estadual Paulista (UNESP). At the State University of Campinas (UNICAMP) she concluded a Master's (1992) degree in Sociology and a Ph.D. (1999) in Social Sciences. She has been Professor at UNICAMP since 1994: first in the Sociology Department and, from 1999 onwards, in the Geography Department. In 2007 she concluded a Post-doc in Geography at the Université de Pau et des Pays de L'Adour in France. Her research is focused on cultural heritage, tourism and the urban environment, and she coordinates the research group Geography, Tourism and Cultural Heritage.
Email: tereza.paes@ige.unicamp.br

Claudia Levy holds a Master's (2006) in Geography from the State University of Campinas (UNICAMP). She completed a Joint International M.Sc. (2008) on Regional Development Planning and Management at the University of Dortmund in Germany and the Kwame Nkrumah University of Kumasi in Ghana under a DAAD scholarship. From 2009 to 2013 she was a Researcher associated with the German Institute for Tropical and Subtropical Agriculture (DITSL), joining the International Center for Development and Decent Work (ICDD) at the University of Kassel in Germany. Within this international network her Ph.D. was completed in 2016 with her project on rural livelihoods and farmer groups in Mozambique.
Email: claudia.levy@ymail.com

Francisco J. Toro is an Associate Professor in the Department of Regional Geographical Analysis and Physical Geography. His doctoral dissertation, completed in 2011, was entitled 'Crisis ecológica y geografía: Planteamientos y propuestas en torno al paradigma ecológico-ambiental'. The main foci of his research publications are theoretical and critical approaches to sustainability and degrowth; the relationship between environmental identity and urban space; and environmentalism within anarchist geographic thinking. He has recently contributed a chapter to *The Radicalization of Pedagogy: Anarchism, Geography, and the Spirit of Revolt* (2016) and is Co-editor of *Historical Geographies of Anarchism: Early Critical Geographers and Contemporary Challenges*.
Email: fjtoro@ugr.es

Zeitfracht Medien GmbH
Ferdinand-Jühlke-Straße 7
99095 Erfurt, Deutschland
produktsicherheit@kolibri360.de